LESSONS LEARNED IN A
BOILER ROOM

Ray Wohlfarth

LESSONS LEARNED IN A BOILER ROOM ver1.03
Copyright © 2011 by Ray Wohlfarth

If you have any questions or comments, please send them to
Ray Wohlfarth
www.FireIceHeat.com
192 Cochran Rd Pittsburgh, PA 15228
Tel 412-343-4110 Fax 412-343-4115
ray@fireiceheat.com

This book is a conglomeration of ideas that I have seen and or used in my career. It is not a design manual. This does not take the place of the boiler manufacturers written instructions, engineering, or code issues that may be in force in your locale. Please follow the boiler manufacturers written instructions that are included in the boiler installation manual. This does not take the place of a properly designed system from an experienced designer.

Thank you for choosing to purchase and read this book. If you find an error, please let me know so that I could change it for the next issue.

Dedication

This book was made possible with the efforts of the following people:

Mr. George Howells

Mr. Paul Lancaster

Mr. Ken Womack

Mr. Ron Lukcic

Mr. Amos Grabe

Mr. Chuck Ray

They allowed me to ply them with questions and patiently answered all my questions. They provided valuable insight and technical expertise. Thank you very much.

I would also like to acknowledge and thank the following companies and organizations for allowing me to complete this book.

ChemWay, Pittsburgh, PA

Engineeringtoolbox.com

Fields Controls

International Code Council

Iron Fireman

Sterling Steam Control Products

Triad Boiler Systems, Inc.

I would love to thank my family for their support, advice and insight

Jon

Ryan

Abby

Conor

Table of Contents

Tales From the Field

Some service call requests will really grab your attention. When the secretary forwarded a call to me where the complaint was that the boiler sounded like a cat in heat, it had my curiosity. I spoke with the maintenance man and he said that every time the boiler started, it sounded like a cat in heat. "What do I do? Here, Listen" he asked as he held the phone out for me to hear. I told him to leave out a saucer of milk and we would be right there.

I accompanied the service technician to the job and sure enough, the copper finned boiler would howl when the burner started. Upon further investigation, we found that the flow switch was bypassed because it would shut off the boiler. The tubes were plugged and had to be cleaned. We cleaned his tubes and he was back to heating that day.

Inside the book, you will find interesting facts that are inside the steam clouds. I think it looks better than saying, "This page left blank intentionally

Did You Know...
There were approximately 400 boiler explosions reported in 1905. The highest total for any year.

Introduction

I have to confess; I am a Boilerholic. I love boilers. They fascinate me. Each one has its own unique personality. I have worked on almost every type of boiler and am in awe of the sheer power that is contained inside each one. My kids hated watching movies with me that featured boilers. We were watching the famous Arnold Schwarzenegger movie, entitled Commando. At the end of the movie, Arnold has a fight scene with the villain inside a boiler room. Inside the boiler room, there is steam flowing everywhere and Arnold rips a piece of pipe attached to the boiler with his bare hands. I yell out "Who in the world is maintaining this boiler? They should be fired." My kids made me leave the room.

If you have the opportunity of designing or installing a replacement boiler room, I hope that this book will be of some assistance to you. I have designed, serviced and installed many replacement boiler rooms in my career. Thankfully, most have worked problem free. The following are rules that I have developed over my 30 years in the industry and hope that they help you. In this book, you will find some redundant information in different chapters. It was my intent for the technician to find the information they require quickly.

Ray's Rules

1 The manufacturer has never heard of the problem that you are having.
- PS: it is a lie….

2 A piece of equipment will always fail 5 minutes after the supply house has closed.

3 The customer farthest from your location will always have a problem on Friday at 4pm.

4 Never plan to repair a condensate pipe leak on a Friday afternoon. You will always replace much more pipe than you anticipated

5 Some clients lie!

6 Out of town or On Line "Experts" are not always experts.

7 Always assume that the old boiler was installed incorrectly.

8 My Favorite…"Behind every successful construction project is a frustrated lawyer"

"It is the owner's intent to have a properly working system regardless of possible omissions or errors."

This is a phrase that is becoming popular in construction documents. What does this mean? It means that the ultimate responsibility for making sure that the system works properly is with you. You will be the one being sued, even if you installed it as it was designed. You will need to check and double check the design.

I have always had a tremendous amount of respect for engineers and designers. When designing a boiler room, they are expected to be experts on every aspect of the job. They have to be able to integrate all the components into one fully operational system. It is a difficult task.

Tales From the Field

We installed new steam boilers for a drug rehabilitation facility. The client called us and said that the boilers would shut off at night, resulting in comfort complaints. We investigated and found nothing wrong with the boilers. We traced the wiring and could find no problems. We thought the boilers could have been wired into the outdoor lights circuit. A data logger, left at the sight, verified that the temperature inside the space did drop each evening about 11:00 pm and rose again around 6:00am. We checked the programming on the thermostats and found no problems there. The thermostat had a locking cover on it as well. We were at a loss as to why the temperature dropped at night. It was discovered that one of the patients in the facility was a HVAC technician that had a key to the locking cover. He would turn the heat down at night to sleep and turned it up when he woke.

Did You Know…
In climate zones that have between 4,000 and 7,000 heating degree days, heating represents about 19% of the energy consumption in commercial buildings?

In climate zones with greater than 7,000 heating degree days, heating represents about 39% of the energy consumption in commercial buildings

Chapter 1
Boiler Room Safety

"Two people hurt in Boiler Explosion. Employees were working on boiler before explosion." Wisn.com

"School 21 Shut by Boiler Explosion and Fire" Yonkerstribunetypepad.com

"Boiler explosion kills five, injures two." Thandian.com

Why would you start a boiler book discussing safety?

A colleague of mine asked me this question as he offered to proof read the book. A boiler room is a very dangerous place. You are surrounded by pressurized vessels that could level an entire building or take off and land hundreds of feet away. In addition to that, you could be exposed to an invisible, odorless gas that could kill you in a matter of minutes, Carbon Monoxide or CO.

The only thing holding this "controlled chaos" in check is a relatively thin piece of metal and a couple safety controls. Rather than frighten you, I hope to instill a sense of respect when servicing a boiler system. Each year, hundreds of people die or are injured due to malfunctioning heating systems.

A sore spot for me is that most maintenance personnel and technicians were never taught about boiler safety. In reality, they simply inherited the equipment. This could lead to a dangerous situation. A friend of mine was a custodian for a school district and one day, the boiler sooted due to a burner malfunction. The custodian opened the boiler and began cleaning the soot from the boiler. His metal brush ignited the soot and the custodian was severely burnt. This tragedy could have been avoided if the custodian had the proper training of the equipment.

On another project, my friend with the gas company asked me to visit one of his clients that was having a difficult time paying their utility bills. He asked me to offer some suggestions to help cut their boiler heating costs. When we walked into the boiler room, wires were hanging out of the burner control panels. As we were talking with the maintenance personal in the boiler room, the lead boiler tech said that he needed to start the boiler. He walked over to the large boiler and opened the door to the base of the boiler. He took a broom, lit the bristles on fire and shoved the lit broom inside. His next task froze me. He opened a 2 ½" manual gas valve and the burner started with a soft "boom." Flames were visible from the opening of the boiler. He pulled the broom out and stomped on it to extinguish it.

I gasped loudly and the maintenance man laughed. *"Why do you do that?"* I asked.
"Aw, the flame safeguard is defective. Do you know how much those things are?" he answered.
"So you bypassed it?" I stammered.
"Yep, been like that for months now. Whenever we need heat, we just light the broom and off we go. You gotta be careful sometimes as that flame will singe your hair." he answered. I do not know whether I was more concerned with how they "lit" their boiler or that they thought this was ok to do that.

Which boiler is more dangerous; water or steam?

When I ask this in my classes, most people feel that steam boilers are more dangerous than hydronic. I can understand their logic. I believe that hydronic boilers are much more dangerous than steam boilers. Steam boilers are designed for steam and have an internal steam chest in the boiler. This allows for expansion of the steam inside the boiler. In addition, the piping could absorb some of the expansion. Hydronic boilers do not have the excess space inside the boiler. As the pressure increases inside the hydronic boiler, it will look for a place to escape. When water turns to steam, it expands at 1,600 its volume, turning the boiler into a rocket. Do an internet search some time for boiler or water heater explosion and see the explosive energy of a water heater or boiler exploding. Mythbusters shows an amazing video on the effects of a water heater explosion on their website.

Did you know?
…that in a recent 10-year period?

- 127 People lost their lives due to boiler & pressure vessel accidents
- 23,000 People were injured due to boiler & pressure vessel accidents.

Who were the people that lost their lives or were injured?
In most instances, it is the people that frequent the boiler rooms like custodians, maintenance personnel, service technicians, and people like you.

Did you know that, according to a study performed by Teledyne Laars, a cast iron boiler rated for 2,000,000 Btuh has the explosive capacity of 32lbs of TNT?

This was from the National Board of Boiler & Pressure Vessel Inspectors…

- Between 1992 and 2003, there were over 30,000 boiler & pressure vessel accidents in the US. 2003 was the last year that the boiler incident report was published by the National Board.
- Cast iron boilers can fracture into many small and/or large pieces in an accident, sending the "shrapnel" in all directions.

What are the top three causes of boiler failures?
1. Operator error, poor or improper maintenance
2. Low water cutoff
3. Failure of Primary Control i.e. Limit control or Relief valve
   - Source National Board

The top cause, operator error and/or poor or improper maintenance could be due to a lack of training or budget cuts. As you know, maintenance is usually the first item to be sacrificed when the money gets tight.

The low water cutoff is the leading mechanical cause of boiler accidents. This again could be attributed to a lack of maintenance or training. The low water cutoff is a safety control that is designed to shut off the burner if the water level drops to an unsafe level inside the boiler. Inside the low water cutoff is either a probe or a float. The low water cutoff should be disassembled and cleaned yearly. The unit should be tested on a regular basis to make sure that it will shut off the burner.

The primary control is a flame safeguard that assures that the burner is operating safely. The limit control will shut off the boiler if the pressure or temperature is too high.

The irony of this is that the majority of these accidents would not happen if the boilers were maintained on a regular basis. The following charts display the causes of boiler accidents that were recorded by the National Board.

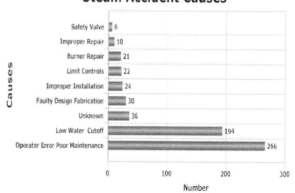

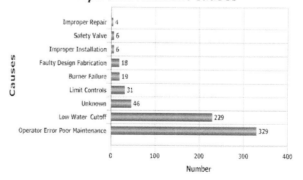

"Plumber faces sentence over fatal boiler explosion"

www.belfasttelegrapgh.co.uk December 7, 2009
Could this be the future? Will the person that does the maintenance be held criminally responsible for a death or injury due to a boiler malfunction? According to the American Bar Association, there are 1,143,358 attorneys in the USA. That breaks down to 22,867 lawyers per state. You really need to document everything and protect yourself.

Hyperbaric Chamber

Carbon Monoxide

Another danger inside the boiler room is carbon monoxide or CO. According to the *Journal of the American Medical Association*, there are over 40,000 visits to the hospital emergency rooms with CO poisoning and over 1,500 people die each year from carbon monoxide poisoning. Carbon monoxide poisoning is the leading cause of accidental poisoning deaths in America. Although most CO poisoning happens during a one-time sudden incidence, it is possible to suffer from chronic CO poisoning. This is when someone is exposed to low levels over weeks or months (for instance, due to a faulty furnace), and experiences symptoms over that time.

Carbon Monoxide is an odorless, tasteless and colorless gas that can kill in minutes. It inhibits the blood's ability to carry oxygen to the body. This lack of oxygen starves your major organs, such as the heart and brain. When CO is inhaled, it combines with the hemoglobin of the blood to form **carboxyhemoglobin (COHb)**. Hemoglobin is used to transport oxygen throughout the body.

Carbon monoxide will replace the oxygen in the hemoglobin. In the presence of carbon monoxide, the hemoglobin cannot transport oxygen.

The levels of carboxyhemoglobin in the body are a factor of the concentration of the gas being inhaled (measured in parts per million or PPM) and the duration of the exposure. What is particularly troubling about Carboxyhemoglobin poisoning is the long half-life. Half-life is a measurement of how quickly the levels return to normal after the exposure to CO has ended. The half-life of carboxyhemoglobin is approximately 5 hours. This means that it will take about 5 hours for the level of carboxyhemoglobin in the blood to drop to half its current level *after* the exposure has ended. For example, if your body has 40% exposure, it would drop to 20% after 5 hours. After another 5 hours, the level would be at 10%. You can see how a prolonged exposure would have a cumulative effect on the body. In some severe instances, people suffering carbon monoxide poisoning have to be placed inside a hyperbaric oxygen chamber to force the CO from the blood.

The following is a chart that details the dangers of carbon monoxide poisoning at various levels. Later in the chapter, we will show the same effects when measured with a standard carbon monoxide detector in PPM.

Carbon monoxide poisoning symptoms

Symptoms of COHb	
% COHb	**Symptoms and Medical Consequences**
15	Mild Headache
25	Nausea and serious headache. Quick recovery after treatment with oxygen and/or fresh air.
30	Symptoms intensify. Potential for long term effects especially in the case of infants, children, the elderly, victims of heart disease and pregnant women.
45	Unconsciousness
50+	Death

Early warning signs of low-level carbon monoxide poisoning include tiredness, headaches, dizziness, nausea or vomiting and shortness of breath. Your skin may also turn pink or red in response to rising blood pressure. If you are feeling any of these symptoms, you may be suffering from carbon monoxide poisoning and should take appropriate action. I recommend the use of a personal carbon monoxide detector when you are working on a heating system. When testing for carbon monoxide in our industry, there are several benchmarks that you need to be aware of. They are listed below.

How much carbon monoxide (CO) is too much?

- 0 ppm Desirable Level
- 9 ppm Maximum allowable concentration for short term exposure in a living area, according to ASHRAE
- 50 ppm Maximum allowable concentration for continuous exposure in an 8 hour period
- 400 ppm "CO Air Free " in boiler flue gas

What is a PPM?

Carbon Monoxide levels are measured in very small amounts called Parts per Million or PPM. To see just how small a PPM is consider this, one part per million equals:

- 1 inch in 16 miles
- 1 Bogey in 3,500 Golf Tournaments
- 1 Square Foot in 23 Acres
- 1 drop Vermouth in 80 Fifths of Gin
- 1 interception in 8,000 football games

How can you test for something that small? I had several relatives that worked in the coal mines. At holidays, they would regale us with stories of life in the mines. One of the most memorable stories was one in which the coal miners used to bring canaries into the coal mine. This was basically a crude carbon monoxide detector. If the canary stopped singing, the coal miners would have to rush from the mine due to elevated carbon monoxide levels. Our industry has come a long way since then. There are very sophisticated detection devices that measure down to a single part per million. They should be part of any service technician's tools. Since it is difficult to measure the levels of Carboxyhemoglobin, you could see the effects of carbon monoxide using the PPM method. Some emission pollutants are measured in Parts per Billion.

Two safety items that I highly recommend

- Install a carbon monoxide detector in every boiler room
- Have a personal CO detector with you any time that you are working in a boiler room.

What happens if you are exposed to CO?

CO PPM	Effects
200 ppm	Slight Headache, Tiredness, Dizziness, nausea after 2-3 hours.
400 ppm	Frontal Headaches 1-2 Hrs, Life Threatening After 3 Hours.
800 ppm	Dizziness, Nausea & Convulsion within 45 minutes. Unconsciousness within 2 hours, Death Within 2-3 Hours.
1,600 ppm	Headache, Dizziness & Nausea within 20 minutes, Death within 1 Hour
3,200 ppm	Headache, Dizziness & Nausea within 5-10 minutes, Death within 30 minutes
6,400 ppm	Headache, Dizziness & Nausea within 1-2 minutes, Death within 10-15 minutes
12,800 ppm	Death Within 1-3 Minutes

When I initially read these effects, my first thought was "Who would have volunteered to be part of this study?"

On a more serious note, the difference between 0 ppm and 12,800 ppm is as little as a ¼" adjustment of the air damper on the burner.

We Have Insurance…

I was talking with a client about a service agreement for his equipment and he replied, "Nope. I do not need a service agreement. I have insurance."

I asked to see a copy of his policy. The following is a transcript of the policy.

> "Accident" shall mean a sudden and accidental breakdown of the Object or a part thereof, which manifests itself at the time of the occurrence by physical damage to the Object that necessitates repair or replacement of the Object or part thereof;
> *but Accident shall not mean*
> *(a) depletion, deterioration, corrosion or erosion of the material;*
> *(b) wear and tear;*
> *(c) leakage at any valve, fitting, shaft seal, gland packing, joint or connection;*
> *(d) the breakdown of any vacuum tube, gas tube or brush;*
> *(e) the breakdown of any electronic computer or electronic data processing equipment;*
> *(f) the breakdown of any structure or foundation supporting the object or any part thereof;*
> *(g) an explosion of gas or unconsumed fuel within furnace of any Object or within the passages from the furnace of said Object to the atmosphere; nor*
> *(h) the functioning of any safety device or protective device.*

I cannot imagine anything that could happen to a boiler that this policy covers. When do you find out what is covered by the insurance policy?
…Usually after an accident.

Have You Ever Had One of "Those" Jobs?

We sold three boilers to a nursing home and we could not get the burners to run correctly. We spent hours on the jobsite and about the same amount of time talking with the factory. They assured us that "We never heard of the problem." The burners would start rumbling and the flue would shake violently. The owner was furious. We had to cut the firing rate drasticallly to keep the boilers running. We were wondering if the boilers would be able to heat the building once the cold weather arrived. You would think that you fixed the problem when they would suddenly start rumbling again. The people living above the boiler room could get no sleep. We were desperate and even contemplated hiring an exorist to remove the demons from the building. Well, not really.

After a couple weeks of this, I received a call from the factory asking me for the address of the problem boilers. I found this curious and asked why. The factory person said that they wanted to look at the building for themselves. The factory said that I did not have to be there. I went to the jobsite myself and found out that there was what they called a "Quiet Recall." They had installed the wrong heads on the burners and had to replace them. Remember Rays' Rule #1.

1 The manufacturer has never heard of the problem that you are having.

     – PS: it is a lie….

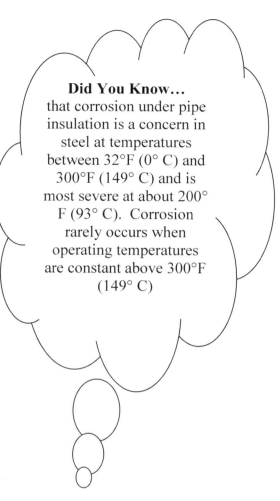

Did You Know...
that corrosion under pipe insulation is a concern in steel at temperatures between 32°F (0° C) and 300°F (149° C) and is most severe at about 200° F (93° C). Corrosion rarely occurs when operating temperatures are constant above 300°F (149° C)

Chapter 2
Boiler Room Walk Through All Boilers

Looks Are Deceiving

I once visited the nicest boiler room that I had ever seen. The floor was cleaned and waxed. There was nothing stacked in the room. It was well lit and the boiler jacket was even spotless. I had entered Boiler Utopia. My initial thought was that nothing could be wrong here. As I walked to the rear of the boiler, I stopped right in my tracks. The outlet piping of the relief valve had a pipe cap installed on it. When I asked the custodian about it, he said, "Yep, that thing leaked all over my clean floor so I fixed it." This could have been one of those statistics in the previous chapter.

System Specialist

While anyone can simply replace a boiler with another one, it is my hope that this book will help you to look at the entire steam or hydronic system. I believe that is what the client wants. I hope that you can offer suggestions to help your client increase comfort and lower their fuel costs. www.heatinghelp.com has some informative books and information for anyone looking to learn more in this industry.

All Boiler Rooms

As a child, I would enjoy watching the Pink Panther Movies, starring Peter Sellers as Inspector Jacques Clouseau. His trusty sidekick, Cato, would wait in hiding to ambush the bumbling inspector. Replacing a boiler in a building where you are not the service contractor could be just like that. There could be many booby traps inside and outside of the boiler room.

Remember Ray's Rule #7 *Always assume that the old boiler was installed incorrectly.*
The most important part of any boiler replacement project is the initial walk through and inspection of the existing system. A mistake or omission in a boiler replacement project could cost thousands of dollars and many headaches. I

have compiled a list of things to look for that may be able to help you to avoid these problems and expenses.

Document Everything

Take pictures of the room and notes about the jobsite. Photos are a great option because they will help you to remember the layout of the boiler room. Some people take videos of the room. I prefer pictures because I could use some of the pictures in my proposal or report. The best time to do your walk through is when the equipment is actually operating. In most replacement projects, you are looking at the boiler room in the summer. If this is the case, perhaps you could ask the owner's representative if you could operate the equipment. In addition, I like to speak with the person that maintains the equipment on a daily basis. They can provide a wealth of valuable information on the actual operating conditions of the equipment.

Invite the boiler representative on your inspection

As the old saying goes, "Two heads are better than one." The representative has most likely seen many more boilers rooms than you have and may help you see potential problems, code violations and opportunities. The representative may also assist you in designing the new boiler room.

Have a copy of the boiler installation codes for your area.

You should always keep a copy of the current boiler codes for your area so that you are installing the boilers according to the current requirements.

Use a Checklist

A checklist is an excellent way to assure yourself that you have looked at all the pertinent areas of the project. I have included a sample one later in the book that may give your some ideas.

Take a Boiler Cut Sheet

Take a copy of the boiler cut sheet with you. This will help you plan how and where the boiler could be installed. The fuel and boiler piping connections may be on the opposite side of where they currently are. It will also help you when estimating the piping, flue and electrical. It may save you an extra trip to the jobsite.

Gas Leak

If you smell a gas leak, the room should be ventilated and all units shut off. If the leak is severe, you should call the fire department. An electronic leak detector is the best choice for explosive leak detection as it is much more sensitive than other options, such as soapy water. Do not use a lighter or a lit match to look for a gas leak! Mercaptan is added to the natural gas in the manufacturing process to allow it to be detected by smell. Natural gas is normally odorless.

Flue Gas Leak

A flue gas leak could be dangerous as well. The flue gases will burn your nose as you breathe inside the room. If you smell flue gases, look for improperly vented equipment. This will be indicated by discoloration on the boiler or water heater or by rust on top of the water heater under the draft diverter. It could be caused by inadequate combustion air venting. Please see the combustion air venting tables later in book. Always remember your carbon monoxide detector.

Are Gas Train Components Vented to the Outside?

Many of the components on a gas train are to be vented to the outside. This is true for gas pressure regulators, gas valves and some gas pressure switches. A common mistake is for the installer to vent all the components in a gas train in the same pipe as the Normally Open Vent Valve. According to UL 795, the normally open vent valve has to be vented separately to a safe location outdoors. No other vents can be combined with it. The normally open vent valve is located between the two electric gas valves in

some gas trains The vents from the regulators and gas pressure switches may be ganged together as long as the cross sectional area is at least the size of the largest vent opening plus 50% of the area of the additional vent pipes. Some new regulators will use a vent limiter that does not have to be vented to the outside.

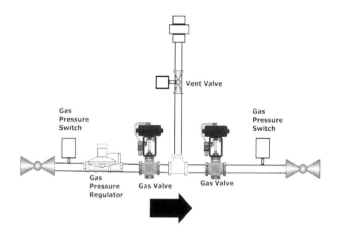

Teflon Tape on Gas Piping

Some manufacturers of boilers, burners and gas valves will void their warranty if Teflon tape or pipe sealant is used on the installation. Please review the installation instructions of the new equipment to see if it mentions the use of Teflon as a thread sealant. If it does, there are other types of pipe thread sealants that can be used that do not contain Teflon.

Teflon Tape

The following is from the Webster Engineering installation manual:

"Warranties are nullified and liability rests solely with installer when evidence of Teflon is found."

Backflow Preventer

Most municipalities require a backflow preventer to be installed on the feed water line to the boiler. A backflow preventer is like a sophisticated check valve. Its job is to prevent the boiler water from mixing with the potable (drinking) water in the building. In many instances, the boiler water could be treated with chemicals that would be dangerous if ingested. This may be required in addition to the building's main backflow preventer. According to section 608.16.2 of the International Plumbing Code, backflow preventers are required where the potable water feeds the boiler.

Combustion Air Openings

Verify the size of the existing combustion air for the equipment room. When you measure the area of the combustion air louver, confirm with the louver manufacturer what the "A_K" factor is for the grill. The A_K factor or Area Factor is the actual free area of the grill. A rule of thumb is that metal grills have about 75% free area and wood grills have about 25% free area. For example, a 10" x 10" louver equals 100 square inches. A metal louver would have 75" free area and a wooden louver would have 25" free area. When calculating sizing requirements, add all the gas fired equipment in the room. This would include the water heater and boilers.

Combustion Air Openings

Each fuel-burning piece of equipment requires combustion air to operate safely. The following are some guidelines to help you see whether the existing combustion air louvers will be adequate for the replacement project. They are from the 2006 International Mechanical Code.

Number of openings required = 2

- Each boiler room should have two openings. One should be within one foot of ceiling and the other opening within one foot of the floor.

Size of Direct Openings

- 1" Free area for each 4,000 Btuh

Horizontal Openings

- 1" Free space for each 2,000 Btuh

Vertical Openings

- 1" Free space for each 4,000 Btuh

Mechanical Ventilation

- 1 cfm per 2,400 Btuh

Louver Screen Size

The International Mechanical Code requires louver screens to be not less than 1/4" and not more than 1/2".

If you have been in our industry for any length of time, you have seen a tradition that has baffled me for years. Why would a plumber, with years of experience, install a water pipe in front of combustion air louver? I finally had the nerve to ask a plumber, who was also a plumbing instructor,

"Why would a plumber that goes to school for 4-5 years to become a master plumber install a water line in front of a combustion air louver where it could freeze?"

He simply smiled and said "Job Security" Watch out for the water line installed in front of the combustion air louver. A boiler room with newer type boilers may be cooler than the older rooms, so freezing could be a concern.

Let us look at our above example and see if the combustion air louver will be adequate for our project. The building has a 10" x 10" opening with metal louvers. That equals 100 square inches. According to our estimates, we have 75% free area or 75 inches free area. The grill is a direct connection to the outside. Based upon 4,000 Btuh per one inch free area, our opening would be adequate for 300,000 Btuh total input. What if our new heating system is larger than that? Well, you have a couple of options. One is to direct vent the combustion air for your boilers. This would involve running duct or plastic piping from the burner combustion air opening to the outside. The other option is to either increase the opening or install a makeup air fan. If you choose mechanical ventilation, you would need 1 cfm per 2,400 Btuh. If your new heating system is rated at 1,000,000 Btuh and your water heater is 40,000 Btuh, you would need a fan capable of delivering 433 cfm. (1,000,000 + 40,000 = 1,040,000 divided by 2,400 CFM = 433 CFM) The makeup air fan should be interlocked with the burner so that the burner does not operate until the airflow is verified. Either of these options could add thousands of dollars to the cost of the project.

In some instances, you may still need more ventilation air to the room even though you have adequate combustion air. Johnston Boiler recommends 2 cfm per boiler HP just for boiler room ventilation. If your new boilers have combustion air directly vented to burner, the room may not get adequate ventilation air and it could get stuffy with elevated levels of CO2 or carbon dioxide. In addition, the boiler room always seems to be the repository of cleaning chemicals and tools. Chemicals should not be stored inside the boiler rooms. The off gassing of the chemicals may mix with the flame and create hazardous gasses.

Is there an exhaust fan?

In most instances, exhaust fans in a boiler room are your enemy. They can pull the flue gases from the boiler or water heater. If this happens, the flue gases would not vent through the vent pipe and chimney. Instead, they would spill flue gases into the boiler room. The flue gases would contain carbon monoxide. In addition, it could cause the boiler to soot. Did you know that only 0.02" W.C. could pull the flame from a water heater or an atmospheric boiler? If an exhaust fan is installed on a project, it should be electrically connected with a makeup air fan to assure that the boiler room does not enter a negative condition. Be careful of air handling units in the boiler room also as they can sometimes cause the room to be negative due to air leaks in the return.

A university discovered this a few years ago. The mechanical room for a dormitory was adjacent to an underground parking lot. The parking garage had a large exhaust fan to void the garage of car exhaust fumes. The garage also had a large make-up air fan to introduce fresh air into the garage. It was a bitter cold day and the internal freeze stat in the makeup air unit tripped, shutting off the fan. Since the exhaust fan and the makeup air unit were not electrically wired together, the exhaust fan continued to operate. The entire garage and the adjoining mechanical room went negative. The boilers and water heater flue gases did not vent and were sent into the dormitory rooms. The facility had to be evacuated. There were more people with white shirts, ties, and clipboards in one place than I had even seen in my life. The blame was placed on the seven-year-old equipment and all of it was replaced. They also installed a safety device that would not allow the exhaust fan to operate without verifying the operation of the make-up air fan. It would alarm if either one did not work without the other.

Pipe Insulation

The International Energy Conservation Code requires pipe insulation on all projects. Please see below. This is crucial on steam systems as the steam could condense faster with a bare pipe than an insulated one. Condensation could cause wet steam. Wet steam increases operating costs and leads to increased maintenance in the system. The wet steam could plug some of the control valves and traps.

Minimum Piping Insulation
International Energy Conservation Code

Fluid	Pipe 1 ½" and below	Pipe > than 1½"
Steam	1 ½"	3"
Hot Water	1"	2"

A Thought on Pipe Insulation

A concern when insulating pipes is corrosion that occurs under the insulation. The insulation could hide and sometimes precipitate the occurrences of corrosion. According to an article entitled, "Inspection Techniques for Detecting Corrosion Under Insulation" by Michael Twomey, corrosion under insulation is caused by the ingress of water into the insulation. The insulation can trap the water like a sponge and hold it in contact with the metal surface. The water can come from rain water, leakage, or sweating from temperature cycling or low temperature operation such as refrigeration units. Corrosion becomes a significant concern in steel at temperatures between 32°F (0° C) and 300°F (149° C) and is most severe at about 200° F (93° C). Corrosion rarely occurs when operating temperatures are constant above 300°F (149° C). Some of the areas that may be susceptible to corrossion under the insulation are:

- Pipes that are exposed to mist overspray from cooling water towers.
- Pipes exposed to steam vents.
- Pipes that experience frequent condensation and re-evaporation of atmospheric moisture.
- Carbon steel piping systems that normally operate in-service above 250° F (120° C) but are intermittent service.
- Deadlegs and attachments that protrude from insulated piping and operate at a temperature different than the active line.
- Vibrating piping systems that have a tendency to inflict damage to insulation

jacketing providing a path for water ingress.
- Steam traced piping systems that may experience tracing leaks, especially at the tubing fittings beneath the insulation.
- Piping systems with deteriorated coatings and/or wrappings.
- Locations where insulation plugs have been removed to permit thickness measurements on insulated piping should receive particular attentions.

Asbestos

Asbestos is another danger inside a boiler room. Inhalation of asbestos causes several serious illnesses including lung cancer, mesothelioma and asbestosis. If the existing heating system was installed prior to 1972, it is a good bet that the system contains asbestos. The asbestos could be on the flue or piping. It was typically used at the pipe fittings. Asbestos looks like a bright white flaky substance. Most facilities have had the insulation tested and should be able to inform you if any asbestos is in the boiler room. Some of the older cast iron boilers used asbestos rope between the sections or on the flue collector to seal in the boiler flue gases. In most instances, the company performing the asbestos testing in the facility would not know that and could have omitted it in their report. If you have any doubt, always have it checked. A cost effective idea is to add a disclaimer to your proposal that asbestos abatement is not covered in your initial quote. The remediation costs could be very high.

Flue & Breeching

If you are reusing the existing breeching and stack, check the new boiler installation manual. It will tell you what type of flue is required for the new boiler. In most instances, the existing flue could only be used with Category 1 type appliances. Before re-using the existing flue, inspect it to see if there are any potential problems. Check and note the following:
- Diameter of horizontal and vertical flues
- Length of flue

- Rise or pitch of flue
- Height of the chimney or stack.
- Construction of existing flue.
- Is chimney lined?

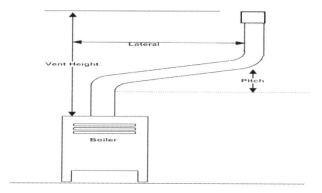

Flue Vent Problems

Flame Rollout

Please notice the discoloration on the boiler. This indicates a dangerous condition. It indicates that the flue gases were not venting through the boiler. Rather, they were rolling out the front of the boiler into the boiler room. Check the draft diverter on the water heater also to look for venting problems. If there is rust on the top of the water heater, this also indicates a venting problem. The cause should be investigated. It may be a leaking air-handling unit in the room, negative conditions in the space, plugged flue passages or inadequate combustion air.

Flue Diameter

Measuring the size of the existing flue is sometimes a challenge, especially a large one. You find yourself squinting and trying to "eyeball" each starting point. An easier way to find the diameter of the flue is to measure the circumference of it using a wire. Then straighten the wire and measure the length. The following table is the circumference for different size round flues. To find the circumference of a circle, the formula is.

Circumference = Diameter x π

To find the diameter, simply divide the circumference by π or 3.1416. When sizing B Vent, the inside diameter is usually one to two inches less, depending on wall thickness. To be sure, check with the vent manufacturer.

Round Flue Diameters in Inches Single Wall			
Diameter	Circumference	Diameter	Circumference
12	37.70	28	87.96
14	43.98	30	94.25
16	50.27	32	100.53
18	56.55	34	106.81
20	62.83	36	113.10
22	69.12	38	119.38
24	75.40	40	125.66
26	81.68		

Insulating the Metal Flue

An installer in our area always insulates the metal flue from a category 1 boiler. This reduces the chances of flue gas condensation in the chimney. It also makes for a more professional project.

Does the Boiler Room Need Heat?

A common occurrence during a boiler room retrofit is that the boiler room may require additional heat. I know that sounds odd. *This is usually discovered after the job is completed.* It could be due to the following:

- The new boilers are more efficient and have a lower jacket loss into the room. If the new boilers are piped in a Primary-Secondary arrangement, the jacket loss will be reduced.

- The combustion air to the room had to be increased to meet the current code.
- The piping is better insulated.

During your walk through, note whether there are heaters in the room. Another often overlooked consideration in a boiler room is the combustion air for the burner. If the boiler room experiences a wide temperature swing, this will affect the fuel to air ratio of the burner. If the burner fuel to air ratio was set when the boiler room temperature was warm, it may have too much air when the room temperature is cooler. For example, if you set the fuel to air ratio when the temperature inside the room is 80^0F, the burner will have 34.6% more air if the temperature inside the boiler room drops to 60^0F due to the density of the air. This will lower the efficiency of the boiler.

Draft Controls

If the building stack or chimney is over 30 feet tall, a draft control is recommended when using Category 1 appliances. This could in-

Barometric Damper

clude anything from a barometric damper to a sequencing draft control. The barometric dampers are more common due to their lower installed cost. When using a barometric damper, the red stops on a barometric damper should be removed if the boiler is firing with only natural gas. The red stops remain if firing with fuel oil. When using a barometric damper, some

Spill Switch

municipalities also require the installation of a spill switch. This will shut off the boiler in the event of a blocked flue or flue gas spillage into the room. Spill switches should be used anytime there is a barometric damper or a draft diverter.

Many of the new boilers and furnaces have integral spill switches.

Sequencing Draft Control

Sequencing draft controls are typically installed in larger facilities, such as hospitals or industrial facilities. They are much more accurate and expensive than a barometric damper. A sequencing draft control consists of an electrically operated damper on the outlet of the boiler, a draft-sensing device and a draft control. The draft control will modulate the draft damper to maintain the desired draft set point. They do require more service more than a barometric damper.

What happens if you do not have a barometric damper? This customer improvised!

Boiler Sooting

Boiler soot is extremely dangerous. It contains fine particulate matter(PM) that could find its way into your lungs and

could cause damage. If you encounter a sooted boiler, consult the boiler manufacturer about how to safely clean it. In most instances, you should wet the soot and use a *nylon* brush to clean it. A warning to you: It goes everywhere! You will need to use protective gear, including a safety mask, gloves and disposable coveralls. The soot will imbed itself into your skin as well, making it very difficult to clean. I was performing a training seminar one day and was discussing soot. One of the attendees told me his secret of how to stay "soot free" when cleaning a sooted boiler. He also informed me that he would have to clean their boilers monthly because of the low water temperature that the boilers were set for. His attire for cleaning the boiler consisted of sealing the sleeves of his shirt with duct tape and covering his face with Vaseline to keep the soot from imbedding itself in his skin. He would wear a tight fitting hat and coveralls. When I asked why he simply did not raise the water temperature, he told me his boss wanted to save money. If the boiler is a cast iron boiler, it is very difficult to clean due to the narrow flue passages. In some severe instances, the sections may need to be disassembled to properly clean the boiler. It is a very messy and dangerous job. If you are using water to clean the soot, be careful about wetting the refractory in the boiler. It could destroy the refractory and you will have to replace it. Adding water to soot could also cause acids to form. It is always less expensive to verify that the burner is operating correctly than to allow it to soot the boiler. Soot can also find its way into the flue piping which could cause severe damage if ignited. When cleaning a sooted boiler, always check the flue as well.

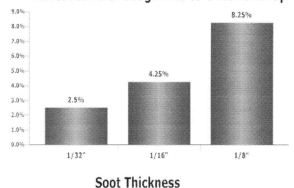

Increased Fuel Usage Due to Soot Build Up

Soot Thickness

Effect of Soot on Fuel Consumption
Soot accumulation on the heating surfaces can dramatically increase fuel consumption. Please see chart.

A Note about Refractory
Refractory inside a boiler is used to reflect the heat and protect the metal surfaces inside the boiler against flame impingement. Always use respiratory equipment to remove or replace the refractory. If you do have to replace some of the refractory inside the boiler, please read the instructions and the MSDS sheets. In most instances, refractory has to be cured and air-dried before heat is applied. If flame is applied to refractory when it is wet, it could cause the refractory to fall apart. The moisture inside the refractory could turn to steam and as we all know, steam will expand at 1,600 times its volume.

Boiler Log
Boiler logs are a great idea for any boiler room. The log contains a list of items that should be inspected on a regular basis on the boiler. These include safety devices on the boiler such as low water cutoff, operating and limit controls and relief valves. They are available on line or through the building owner's insurance company.

Electric Availability
When replacing an older, atmospheric boiler with a new one with a power burner, you will need to find additional electricity. A cut sheet on the new equipment would help you to see what is required.
- ✓ How many breakers and what amperage are the available breakers?
- ✓ Will the new boiler(s) be single or 3 phase?
- ✓ Are replacement breakers available for the old electrical panel?

You will need the following information from your walk through:
- ✓ Manufacturer of existing electrical panel(s)
- ✓ Model of panel, breakers

- ✓ Is 3 phase available? (If new boilers are 3 phase)
- ✓ What is the voltage on site, 208, 480, etc.?
- ✓ How far will you have to run the wiring?
- ✓ Does new wiring have to be in conduit?
- ✓ What size is existing wire?

Another consideration is the sensor wiring for the boiler control panel. In most instances, it cannot be installed in the same conduit as the line voltage. The higher voltage may affect the sensor reading.

Scale Build Up

Scale formation on the waterside of the boiler could dramatically increase the operating costs. The following chart below illustrates the estimated costs of scale formation. Scale forms usually from an increase in makeup water due to a leak in the system or if the building has hard water. Scale will sound like Rice Kripies Cereal i.e. Snap, Crackle, Pop when the flame is on.

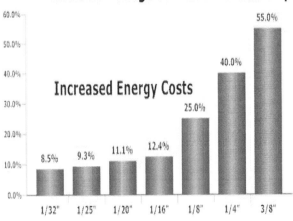

Increased Fuel Usage Due to Scale Build Up

Gas Supply

This could be an expensive area if a mistake is made here. Some of the things to check are:

- • Size of the gas pipe feeding existing boiler room and each boiler
- • Existing gas pressure

An old, atmospheric boiler may have been firing at only a few inches of water column of gas pressure, typically 3" to 5" w.c of gas pressure. The new, power burners will use up to 14" gas pressure for commercial buildings and sometimes more than that. This is four times the gas pressure than the older atmospheric boilers needed.

If the existing boilers used high-pressure gas, the new boilers will have to be ordered with a different gas train. In some instances, a new high-pressure regulator would be required. Most new burners have a standard of 14" W.C. or ½ pound of gas pressure. If the building has high gas pressure (over one pound pressure), a regulator may have to be installed to reduce the gas pressure. Excessive gas pressure can destroy the standard gas pressure regulators. The cost of a new high-pressure regulator could be anywhere from several hundred to several thousand dollars. Note the pipe diameter that feeds the old boilers and the main pipe size entering the boiler room. For example, if the old 1,000,000 Btuh boiler had a ¾" gas pipe feeding it, the piping was probably undersized. An undersized gas pipe could also indicate that the existing boilers were not firing to capacity. The boiler, rated at 1,000,000 Btuh, may have only been firing at half its capacity. This may help you in the sizing of the new heating system. Clocking the gas meter would verify the actual firing rate of the existing heating system. That is covered later in the book. If the under fired boiler successfully heated the building, you may be able to resize your new boiler.

Some Notes on Gas Piping

- ✓ Gas fittings should be malleable iron and not black iron.
- ✓ Many gas companies do not permit bushings to be installed in gas lines. They prefer bell reducers.
- ✓ Threaded fittings greater than 4" shall not be used except where approved. IFGC 403.10.4
- ✓ Piping shall not be installed in or through a ducted supply, return or exhaust, or a clothes chute, chimney or gas vent,

dumbwaiter or elevator shaft. IFGC 404.1

✓ Piping in concealed locations shall not have unions, tubing fittings, right and left couplings, bushings, compression couplings and swing joints made by a combination of fittings IFGC 404.3

IFGC = International Fuel Gas Code

Gas Pipe Threads

The following table shows the length and approximate number of threads on gas lines.

Iron Pipe Size (Inches)	Approximate Length of Threaded Portion	Approximate number of threads to be cut
1/2	¾	10
¾	¾	10
1	7/8	10
1 ¼	1	11
1 ½	1	11
2	1	11
2 1/2	1 ½	12
3	1 ½	12
4	1 5/8	13

Gas Meter

As a meter ages, it sometimes does not register the proper amount of gas consumed by the facility. It may be inadequate to meet the pressure requirements of the new equipment. In many cases, the gas company will require replacement of meter when performing a large project like a boiler replacement. In most instances, the gas company will install the new meter for no charge. This may be a rather large expense if the gas company will not pay for it.

Which is Stronger? Styrofoam or Steel?
We once sold some boilers to a school and could not get the gas pressure we needed to properly fire the boilers. The gas pressure was high enough outside the building but dropped once it entered the building. There was finger pointing and threats of lawsuits. A meeting was called that included the owner's representatives, the gas company, the installer and us. The collective group decided to connect high-pressure air to the pipe. There was a sudden whoosh and a loud bang. We found a crumpled coffee cup that someone had placed inside the gas line.

Sizing Corrugated Stainless Steel Tubing (CSST) Natural Gas

Inlet Pressure	Pressure Drop	Specific Gravity
Less than 2 psi	0.6"	0.60

Tube Size	Length				
EHD	5	10	15	20	25
	Capacity in Cu ft per Hour				
13	46	32	25	22	19
15	63	44	35	31	27
18	115	82	66	58	52
19	134	95	77	67	60
23	225	161	132	116	104
25	270	192	157	137	122
30	471	330	267	231	206
31	546	383	310	269	240
37	895	639	524	456	409
46	1792	1260	1030	888	793
48	2070	1470	1200	1050	936
60	3660	2930	2400	2080	1860
EHD	30	40	50	60	70
	Capacity in Cu ft per Hour				
13	18	15	13	12	11
15	25	21	19	17	16
18	47	41	37	34	31
19	55	47	42	38	36
23	96	83	75	68	63
25	112	97	87	80	74
30	188	162	144	131	121
31	218	188	168	153	141
37	374	325	292	267	248
46	723	625	559	509	471
48	856	742	665	608	563
60	1520	1320	1180	1080	1000

EHD = Equivalent Hydraulic Diameter
Each cubic foot of gas equals 1,000 Btuh

CSST Sizing
Rules of Thumb for EHD Sizing

EHD	Pipe Size	EHD	Pipe Size
15	3/8"	37	1 1/4"
19	½"	46	1 1/2"
25	3/4"	62	2"
31	1"		
Verify with manufacturer			

Gas Pipe Line Sizing

	Pipe Length			
Steel Pipe Size	10 Feet	20 Feet	40 feet	80 Feet
	Capacity in Cubic Feet per hour			
½"	120	85	60	42
¾"	272	192	136	96
1"	547	387	273	193
11/4"	1,200	849	600	424
1 ½"	1,860	1,316	930	658
2"	3,759	2,658	1,880	1,330
2 ½"	6,169	4,362	3,084	2,189
4"	23,479	16,602	11,740	8,301

Each cubic foot of gas roughly equals 1,000 Btuh

What is Stored in the Boiler Room?

In many facilities, the boiler room becomes a storeroom for anything. This could lead to a dangerous condition. A school in our area stored papers next to the boilers. The flames rolled out and caught the papers on fire, damaging the boiler and the boiler room. Most boiler manufacturers require a minimum clearance around the boiler. The International Mechanical Code calls for 18" around the boilers. This is in case there is flame roll out of the boiler. Gasoline, cleaning chemicals or refrigerant are also dangerous to store in the boiler room.

Water Meter on Makeup Pipe

A water meter installed on the makeup water line is a cost effective idea. This will allow the owner to monitor makeup water to the system. Excessive makeup water could indicate a leak. Fresh water introduced into the system can lead to premature failure of the boiler and the piping. It could also increase the water treatment costs. The best water treatment is a tight, non-leaking system. The makeup water should be monitored as part of the preventive maintenance service. It

should be checked monthly. It is common for steam systems to lose some water because of leaking steam traps, leaking air vents and the boiler feed unit vent. The boiler pressure also factors in to the amount of water lost. For example, if the steam pressure is high, the condensate could flash to steam and be lost through the vent on the condensate tank.

Installation & Removal of the Heating System

If you are replacing a cast iron boiler, the boiler can be disassembled and the sections removed through most standard doors. However, the sections are going to be very heavy. Each section may weigh several hundred pounds. A steel boiler may have to be cut up and carried out in pieces. In any event, it will be heavy work.

Some questions to consider:
- Will the new boiler fit through the existing doorway?
- What is the door opening size?
- Are there steps to get into the boiler room?
- Is there an elevator? If so, will the client allow you to use the elevator? Will it be able to handle the weight of the existing or new boilers?

When installing a large horizontal fire tube boiler, it may not fit through the existing door openings. Consideration should also be given to the service area for the boiler. The horizontal fire tube boiler will require enough room to change the tubes. The service area 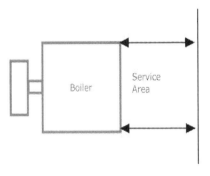 required may be the same length as the boiler. For example, if the boiler tubes are seven feet long, you will need at least that much room either behind or in front of the boiler for tube replacement. Many boiler rooms do not have that much room. An option that we have seen in some boiler rooms is to install a door in front of or behind the boiler that can be opened to allow future tube replacement.

The demolition will be extremely dirty. The old boiler will have dirt, soot, and black water inside. This will be tracked throughout the building without precautions. Carpet protectors should be covering the floors to the outside and the bathrooms. The carpet protectors are available at most carpet stores and big box stores. Shoe covers are also a good way to reduce dirt tracked into the building.

Emergency Door Switch

An emergency door switch should be installed at every exit from the boiler room. This is part of the ASME Code CSD1, which is code for many states. Even if your state has not adopted the code, this is a great safety item. It should be wired into the boiler safety circuit to shut off the boilers in the event that the switch is disengaged. The switch should be on the outside of the boiler room so that the boilers can be shut off without entering the boiler room in the event of a malfunction. In some schools or other buildings where the occupants could have access to the switch, you may want to install it just inside the door. With any of the switches below, you may want to wire in an audible and/or visual alarm that will notify you when the switch is turned off or the button is depressed.

What is the purpose of installing a door switch?
There is a strange noise coming from the boiler room. You open the door and are greeted with a wall of hot, wet fog. It pours from the doorway and trips the smoke detector in the hall. The boilers are running wild inside the room and must be shut off. There is a smell like burning wire as you enter the room. There is a banging sound coming from inside the boiler room. You slowly walk across the boiler room, hoping to remember the piping layout. The light bulb in the room suddenly pops from the wet steam hitting it, startling you. "What would have happened if the room was full of natural gas?" you ask yourself. Your flashlight does not penetrate the mist. You

are, for all intents, blind. You walk toward the noise, waving your hands in front of you liked a crazed ninja warrior, to prevent walking into a pipe. Then you wonder "What if the room is filled with carbon monoxide?" Should you hold your breath? You cannot smell or taste anything but then you remember that carbon monoxide is odorless and colorless. Your pace quickens and your heart is racing. It feels as if you have been here for an hour. Your ankle connects with a pipe and the pain races up your leg. You curse to yourself and continue toward the noise and find yourself in front of a burner. "Where is the switch?" you ask yourself. You feel all over the burner and push everything that feels like a switch. The burner finally shuts off and then you have to find and shut off the other boiler. You slowly manage to shut off the other boiler and sigh deeply. It then dawns on you that you now have to make the same trip in reverse to escape. After tripping over a couple pipes and banging your head on the low ductwork, you realize the value of an Emergency Door Switch

A simple door switch like the one to the right could work. One word of caution is that it could be shut off when the person leaves the room, thinking it is a light switch.	
This type of push button switch will work but some people have bumped it by mistake on the way out of the room and shut off the equipment.	
This is a common type of emergency shutoff switch.	

Gas Train

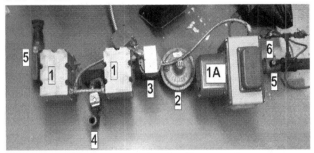

The gas train is comprised of several components in the gas piping for the boiler. The main components in the gas train include the following:

1-Safety Shutoff Valves
1A – Electric Valve Only shuts off when boiler temperature is too high
2-Gas Pressure Regulator
3-Low Gas Pressure Switch
4-Vent Valve
5-Manual Gas Valve(s)
6-Pilot Takeoff

Manual Gas Valve

In most applications, there are two manual gas valves. One is upstream of all components. This valve usually has a tapping for the pilot line. That is the one pictured at right.
Please note the gas flow when installing this valve. The pilot tapping is upstream of the main gas valve. This will allow you to perform tests of the flame safeguard. There is another valve that is downstream of all other components. This will allow you to isolate the gas train components.

Electric Gas Valve

This is the most important part of the gas train. When the safety controls are satisfied, voltage is applied to the gas valve, opening it. Check the following items:

- Wiring connections
- Gas Leaks
- Hydraulic Fluid Leaks (usually visible in valve window)

Some solenoid valves use an adjustable bleed fitting on the vent connection. This turns the normally On Off valve into a slow opening valve. This looks like an

Adjustable Oriface

ordinary brass ¼" fitting that is used to vent the valve but it has an adjustment screw that controls the speed of the valve opening. If you are having problems with rumbling, this device may solve your problem.

Gas Pressure Regulator

As the name implies, this device on the gas train will regulate the gas pressure to the burner. Some of the new types of regulators feature a limiter, which does not require venting to the outside. On smaller boilers, you would typically see a range of 3-7" w.c. (w.c. = water column) Larger commercial boilers will usually see a range of 7 to 14" w.c. On smaller, residential boilers, the regulator is part of the gas valve. The gas pressure regulator must be installed up stream of or before the electric gas valves. If not, the burner will start with a loud bang until the regulator controls the gas pressure to the proper setting. (See picture) It is also a good

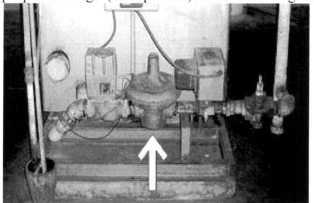

idea to have a straight piece of pipe on the outlet of the regulator that is ten pipe diameters long. For example, if you are installing a 1" regulator, you should have a 10" nipple downstream of the regulator. This helps to reduce the turbulence and the chances of noisy combustion. The regulator may require venting to the outside. When venting the gas train components outdoors, you should consider installing an insect guard on the discharge. This may be something as simple as a screen. Insects will sometimes block the outside termination of the vents. If the vent becomes plugged, the boiler will not fire. The vents cannot be combined in the same pipe with the normally open vent valve on a block, block and bleed gas train. Some regulators allow you to change the range of the regulator by switching the spring inside the cap.

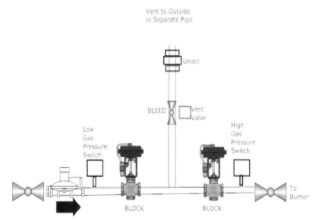

Block, Block and Bleed Gas Train

The normally open vent valve is installed to vent any gas that may leak past the first valve on the gas train when the burner is not firing. The

sequence of operation for this type of gas train is as follows:

Upon a call for heat, the burner ignites the pilot. Once the pilot operation is verified, the two block valves open and the normally open vent (Bleed) valve closes. When the call for heat is satisfied, the block valves close and the bleed or vent valve opens. A union should be installed above the vent valve to allow inspection of the valve and to facilitate leak testing of the valve. A client called me to say that his fuel costs went up drastically one month and asked us to see if we could see a problem. His maintenance man had replaced the normally open vent valve with a normally closed one. Anytime that the burner was on, the vent valve was open spilling natural gas from the 1" pipe. His gas bills dropped dramatically when it was replaced with the right type of valve.

Two Gas Valves? Huh!

A contractor that had never installed a commercial boiler project installed a boiler for his church. He called to inform us that the job was ready to be started. He was anxious to get paid. When we arrived at the project, I looked at the gas train and was shocked. The first gas valve was missing. Luckily, the manual gas valve was shut off or else gas would have poured from the normally open vent valve. I asked what happened to the gas valve and he answered, "I thought they sent an extra one."

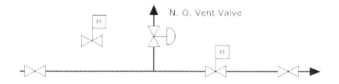

Gas Pressure Switch

The gas pressure switch will shut off the burner in the event that the gas pressure is either too low or high. There is usually a manual reset button on the unit that will need to be depressed in the event of a "trip" Most of the older style gas pressure switches require venting to the outside. On large boilers, there may be two gas pressure switches. One is a low gas pressure switch and the other is a high gas pressure switch. The high gas pressure switch will be located between the last electric gas valve and the burner. It is usually on the burner head. The low gas pressure switch will be located on the gas train, between the gas pressure regulator and the first gas electric gas valve. The low gas pressure switch will shut off the burner if the gas pressure to the boiler is too low. The high gas pressure switch will shut off the burner if the gas pressure is too high. I know, Duh! We use no imagination in naming components in our industry. You may remember reading when a main gas pressure regulator malfunctioned in a Chicago suburb several years ago. People were reporting six-foot high pilot lights on their stoves.

Low and High Press Gas Cut-off

Reuse Existing Gas Train Components

This is not something that I would recommend doing. Each burner is designed with a certain style and type of component engineered for the equipment. If the old components were used on the new burner, the warranty could be voided.

Is There an AC Unit in the Boiler Room?

If your new boiler is located in a room with an air conditioning unit, there is a code covering that application. According to the 2006 International Mechanical Code:

- 1105.5 Fuel-burning appliances. Fuel-burning appliances and equipment having open flames and that use combustion air from the machinery room shall not be installed in a machinery room.
- Exceptions:
- I. Where the refrigerant is carbon dioxide or water.
- 2. Fuel-burning appliances shall not be prohibited in the same machinery room with refrigerant-containing equipment or appliances where combustion air is ducted from outside the machinery room and sealed in such a manner as to prevent any refrigerant leakage from entering the combustion chamber or where a refrigerant vapor detector is employed to automatically shut off the combustion process in the event of refrigerant leakage.

In other words, the combustion air for each boiler has to be introduced from the outside. An atmospheric boiler or water heater will not meet the code. Another consideration would be to install a boiler with a sealed combustion design. This code also applies to a water heater. The other option is to install a refrigerant monitoring system. Note: If you are using a direct-ducted system for the combustion air, it must be sized properly. If the burner start-up is performed in very cold weather, slightly more air may be required to compensate for the air density once the outside temperature rises. If not, this could lead to elevated carbon monoxide readings and a sooted boiler.

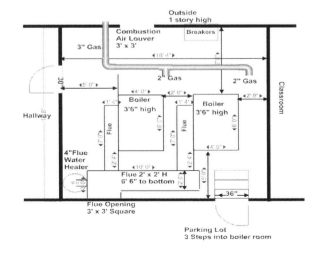

Boiler Room Drawing

A boiler room plan drawing with all the components is an excellent resource for your files. The drawing should include the following:

- Existing boiler location
- Boiler piping
- Access to the boiler room
- Location of combustion air louvers
- All pertinent components, such as water heater location

You want to note items on the drawing that may be helpful in your pricing. This may include electrical panels, ceiling height and existing breeching and piping height. The drawing will also assist in developing a plan for the removal of the old boiler and the installation of the new boilers. It will allow you to see if your proposed heating system will have the proper clearances. In addition, some municipalities require a ¼" scale drawing to get a permit for your boiler replacement permit. Another item to note is how to move the equipment through the room. For example, whether you can use a pallet jack or overhead crane. Look at the drawing above, you will see that the door to the outside is 36" This may limit the size of the boilers that will fit through the door.

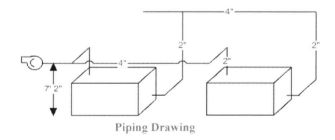

Piping Drawing

Piping Diagram

A piping diagram is also an excellent reference guide. It should illustrate how the existing boilers are currently piped as well as the pipe sizes. In the piping diagram above, you will see that a hydronic heating system piped like that would not meet the current International Energy Conservation Code. Please see the code pertaining to this in the International Energy Conservation Code.

"Boiler plants including more than one boiler shall have the capability to reduce flow automatically through the boiler plant when a boiler is shut down."

To meet the code, you would have to isolate the unfired boiler. To do this, either install automatic isolation valves or re-pipe the header as a primary secondary main. This could be very expensive if it was not included in the price.

When using isolation valves, there are a couple factors to consider. The first is the flow through the boiler that is firing. When one boiler is isolated, all the flow would go through the boiler that is firing. Verify that the flow is not excessive for the boiler design. Excessive flow could void the warranty and in addition, could increase fuel consumption.

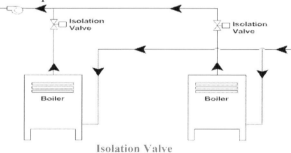

Isolation Valve

Consider the drawing above. If both valves should close, the pump will be starved for flow. The closing time of the isolation valves could be another area of concern. If the valve closes immediately after the firing cycle, there is still heat being generated inside the boiler by the internal hot surfaces. Without flow, this heat build-up could cause the manual reset temperature limit control to trip, meaning no heat on the next call for heat. A time delay relay may be required so that valves do not close too quickly. The last item to consider is that the isolation valve should have an end switch that will prevent the operation of the burner until the valve is fully open. This will limit the possibility of inadequate flow to the boiler. The following is a primary secondary piping arrangement. Most boiler manufacturers are piping their boilers using this strategy. It has several benefits and will reduce operating costs. See how the boilers can be isolated from the main system piping. When piping the boilers in a primary secondary arrangement, the supply to and return from each boiler must be within 12" of each other. If they are further apart, migratory flow could occur through the isolated boilers.

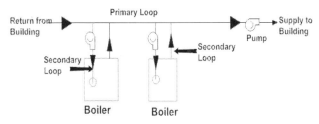

Primary Secondary Piping

Pipe Sizing

Pipe sizing can dramatically affect the success of your project. If the piping is too small, you could have a very upset client due to a lack of heat and high fuel costs. The following is a list of the outside diameters of common pipes found in a boiler room. If unsure about the pipe size, wrap a wire around the pipe, unfold it and then measure the length. This will be the circumference of the pipe.

Standard Black Iron Piping Schedule 40		
Pipe Size	Outside Diameter (O.D.)	Circumference
¼ "	0.540"	1.696"
3/8"	0.675"	2.121"
½"	0.840'	2.639"
¾"	1.050"	3.299"
1"	1.315"	4.131"
1 ¼"	1.660"	5.215"
1 ½"	1.900"	5.969"
2"	2.375"	7.461"
2 1/2"	2.875"	9.032"
3"	3.500"	10.995"
4"	4.500"	14.137"
5"	5.563"	17.476"
6"	6.625"	20.812"
8"	8.625"	27.095"
10"	10.750"	33.771"
12"	12.750"	40.054"

Common Fraction to Decimal				
Fraction	Decimal		Fraction	Decimal
1/16	0.0625		9/16	0.5625
1/8	0.125		5/8	0.625
3/16	0.1875		11/16	0.6875
1/4	0.250		3/4	0.750
5/16	0.3125		13/16	0.8125
3/8	0.375		7/8	0.875
7/16	0.4375		15/16	0.9375
1/2	0.50		1	1.00

Standard Copper Tubing Type K,L,M		
Pipe Size	Outside Diameter (O.D.)	Circumference
½"	0.625"	1.964"
¾"	0.875"	2.749"
1"	1.125"	3.534"
1 ¼"	1.375"	4.319"
1 ½"	1.625"	5.105"
2"	2.125"	6.675"
2 1/2"	2.625"	8.246"
3"	3.125"	9.817"
4"	4.125"	12.959"
6"	6.127"	12.248"
8"	8.125"	25.525"
10"	10.125"	31.808"
12"	12.750"	40.054"

What if the existing boiler has no information?

It is always a good idea to know what is currently installed and compare it to what you are proposing. What if the current boiler has no data tag?

Pipe Sizing Measure the supply and return piping to estimate the flow. Look on the fittings to see if the pipe size is stamped on it.

Flue Sizing The flue size may give you some idea of the sizing as well.

Measure the Boiler If you get the actual physical measurements of the existing boiler, you may be able to look on the manufacturers website for some archived information. If not, perhaps the manufacturer's representative could provide you with some information.

Pump GPM The pump gpm may give you an idea of the capacity of the existing equipment.

Count The Burners If the boiler uses ribbon burners, you may be able to count them and estimate the flow. A rule of thumb is that ribbon burners will have a capacity of 100,000 Btu/Hr each.

Clock the Burner This is the most accurate way to verify the firing rate of the burner.

All of these together may give you an idea of the size of the existing boiler.

Water Treatment

When installing a new system, provisions for water treatment should be included. A competent water treatment specialist could be a valuable ally for you. Without proper water treatment, the new boilers will have a dramatically lower life. This will reflect on your reputation.

A director of maintenance for a local college purchased some boiler chemicals from the sales person that sold him soap. *I know what you are thinking, "What does a soap salesman know about boilers?"* The salesperson told him that he would save "lots of money" with this product because it would rid his system of scale in the sixty-year-old piping and the thirty-year-old steam boiler. *You already know where this is going.* Shortly after installing the new "miracle" chemical, leaks started to develop everywhere. The piping system turned into a sprinkler system. *This was an added benefit that the sales person forgot to mention.* The "miracle" chemical caused the college replace his boilers with four of mine and about a mile of condensate pipe. *Overall, I endorse the stuff.* A qualified water treatment professional is a great person to know.

Water Treatment Feeder

Pot feeders (See Picture) are traditionally used on hydronic systems. The pot feeder is usually a sidearm type feeder that has isolation valves. The pot feeder could also include a filter that will strain the system water. If using a filter as well, a flow indicator should be installed to indicate whether the filter is plugged and should be cleaned or changed.

On steam systems, several chemicals may be introduced into the system. Some may be injected into the boiler feed unit and some may be directly injected into the steam supply piping. Makeup Water, Oxygen and Carbon Dioxide are the three main enemies of a steam system. Makeup water introduces hardness, oxygen and carbon dioxide into the system. Oxygen molecules will pit the inside of the boiler or piping causing a leak. Carbon Dioxide forms Carbonic Acid inside the steam system. Carbonic acid occurs when carbon dioxide is mixed with water. This is usually noticed on the condensate piping.

Makeup water will bring in fresh, untreated water. This fresh water contains minerals such as calcium and magnesium. This is usually referred to as "Hardness" These minerals will deposit themselves on the heating surface and act as an insulator. This reduces the heat transfer ability of the boiler as well as the efficiency. In severe instances, it could destroy the boiler. Hardness inside a steam boiler should be limited. One way to limit the hardness is to install a water softener. A softener also reduces the amount of water treatment chemicals that are used.

Blow down is crucial on steam systems and provisions should be made for a blow down valve and / or a surface skim tapping. Location of the blow down fitting is critical as dissolved solids will generally collect at 4-6" below the water level. The skim tapping is typically just below the normal boiler operating level. Some systems require automatic blow down of the boilers. In this case, the water treatment expert could inform you of this. Also, check your installation codes to see if the blow down piping can be piped directly to the drain. Many municipalities do not allow that because the high temperature boiler water could melt the PVC drainpipe in the building. PVC piping is limited to 140°F. In that case, you would need a blow down cooler. This device reduces the blow down discharge temperature to allow safe draining of the water by injecting city water into the unit.

To limit oxygen pitting in a steam system, a deaerator is often installed. This will remove most of the oxygen in the feed water to the boiler, limiting the damage from oxygen pitting. In addition, the customer may use some sort of oxygen scavenger water treatment chemical to remove more of the dissolved oxygen. Oxygen pitting is an amazing process. If you look at a nipple that was attacked by oxygen pitting, the oxygen molecules will attack the same spot over and over again until it develops a hole. It almost looks like someone drilled a hole in the nipple. The rest of the nipple could be relatively damage free. It is like a shark that is attracted to the scent of blood.

Carbonic acid will wash away the bottom of the nipple or piping. The top of the piping may be the same thickness as when it was installed.

Some installers use a sacrificial anode inside the boiler feed tank. A sacrificial anode is a rod that is made of a softer metal like magnesium. The theory is that the aggressive water will attack the softer metal first and not the boiler piping or tubes. Sacrificial anodes are commonly installed in water heaters also.

Remember Ray's Rule *4 Never plan to repair a condensate pipe leak on a Friday afternoon.*
I have seen this on many projects. You see a nipple leaking on the condensate line and think to yourself that it will be only a few minutes to change the leaking nipple. You attach a pipe wrench to the pipe and get that sick feeling in your stomach as the nipple collapses. As you inspect the nipple, you see that it is paper-thin. If that is the case, the adjoining nipples and pipe are probably in the same condition. You typically have to go back to the closest fitting and replace the piping. It is not something that I would attack on a Friday afternoon. This is usually caused by carbonic acid.

Should dye be used in the water treatment?
Colored dye is sometimes included with the water treatment as a way to visually display if the system piping is leaking. One hospital in our area found the value of the dye in their treatment. They had a stainless steel water fountain installed when renovating one of their wings. The fountain was in service for several months without incident until one day, a nurse filled a white Styrofoam cup with water from the fountain. She saw that the water had a pink hue. The nurse called the maintenance department and asked why the water was pink. After some investigation, it was discovered that the installer had connected the fountain to the chilled water loop. Without the dye, the mistake could have never been found. Many more people could have unknowingly ingested the water treatment chemicals.

Feeding the Chemical Treatment

Chemical Feed Pump

Chemical treatment for steam systems should be proportiontely fed into a heating system. Typically, the water treatment company may use an injection pump for each chemical. The pump feeds a small amount of chemical at a time. Some facilities will manually feed the chemical into the boiler feed tank if their reading is low. This is referred to as "slug feeding" There are several drawbacks to this type of chemical feeding. First of all, the employees would be subjected to having the chemical spill on them when pouring it into the boiler feed tank. In addition, the system will experience wide variations of the chemical levels inside the boilers. This could lead to bouncing water levels and potential boiler problems. The other disadvantage is that there will be a disproportionate amount of chemicals in each boiler. The lead boiler on the day that the chemicals are fed will receive most of the chemicals. The lag boilers will get a substantial amount less or even none at all.

Water Treatment Information

Carry-Over: It is the continual discharge of impurities with steam. It can be detected by checking the conductivity of the condensate water.

Caustic Embrittlement: It is the weakening of the steel because of inner crystalline cracks. It is caused by long exposure to highly alkaline water and or stress of the metal.

Corrosion: It is the result of low-alkaline boiler water or the presence of free oxygen or both.

Foaming: It is a layer of foam on the surface of the water. It is most commonly caused by oil and other impurities. It may appear as a film atop the water, which impedes the steam bubbles from breaking through.

Priming: It is when large slugs of water are suddenly discharged from the boiler with the steam. This is caused by impurities in the water and boiler design.

Scale: It is a deposit of solids that form on the heating surfaces.

TDS: Total Dissolved Solids

Typical Steam Boiler Water Treatment Normal Levels
TDS 1,500-3,000 ppm or 2,000-4,000 micromhos or microsiemens
Microsiemens x .70 = TDS
Phosphate 30-60 ppm
Hydroxyl alkalinity 200-400 ppm
Sulfite 30-60 ppm SO3
Boiler pH 7-9* 0r 9.5-11(*pH over 11 can cause boiler foaming in some boilers.*)
Some boilers prefer 10-11.0 pH
Condensate pH 8.2-9.0

*Each boiler is different. Please check with manufacturer as to their requirements. The above are just some industry rules of thumb.

Steam Boiler and Bouncing Water Levels
We were asked to consult on a steam boiler project that was having numerous problems. The boiler water level would bounce greatly and trip the low water cutoffs. Some areas of the building would flood and overcome the steam traps, losing heat. After some testing, we found that the water inside the boiler contained elevated total dissolved solids or TDS. The boiler had carryover, which meant that it was taking water from the boiler and sending it into the steam piping. The carryover was caused by incorrect installation of the near boiler piping. The chemical treatment was an all-inclusive treatment that contained an oxygen scavenger as well as several other chemicals to reduce scale and erosion. The oxygen scavenger removes the oxygen in the boiler to avoid oxygen pitting. When the water from the boiler was carried into the steam and return piping, the air inside the return piping would consume the oxygen scavenger. When the maintenance man checked the oxygen scavenger level in the boilers, it would read low. He would then add more chemicals. This led to overfeeding of the chemicals, elevated TDS, bouncing boilers and no heat because the other chemicals in the treatment system were not depleted. To resolve the problem, the near boiler piping had to be replaced. It was an expensive repair.

Water Treatment Equations

Cycles of Concentration =
$$\frac{Chlorides\ or\ TDS\ in\ Boiler\ Water}{Chlorides\ or\ TDS\ in\ Feed\ water}$$

Boiler Blow down = 4 to 8 % of boiler makeup

One Cubic foot of water = 7.5 Gallons

$$\% \text{ of Blowdown} = \frac{Chlorides\ or\ TDS\ in\ Feedwater}{Chlorides\ or\ TDS\ in\ Boiler\ Water}$$

$$\% \text{ of Makeup} = \frac{Chlorides\ or\ TDS\ in\ Feedwater}{Chlorides\ or\ TDS\ in\ Make-Up\ Water}$$

Pounds of Chemical/1,000 Gallons Water =

$\frac{PPM\ of\ Product}{120*}$

*120,000 gallons of water weighs about 1,000,000 pounds.

A Note on pH

A good pH tester is a good service tool when diagnosing steam boilers. Each boiler manufacturer will require different pH levels inside their boiler. For instance, most steam boilers require a pH level of 7-9 while some manufacturers prefer 10-11.5 Levels that high could cause foaming and bouncing water levels in the boiler designed for 7-9. When measuring pH, a level of 7 is considered neutral. Readings below 7 are acidic. Readings above 7 are basic.

What some people do not realize is that pH readings are logarithmic. For instance, a reading of 6 is **Ten** times more acidic than a reading of 7 and a reading of 5 is **Ten** times more acidic than a reading of 6. A reading of 5 is **One Hundred** times more acidic than a reading of 7. That is why a proper pH reading is critical on boilers

pH Scale

10,000,000	PH=0	Battery Acid
1,000,000	PH = 1	Hydrochloric Acid
100,000	PH = 2	Lemon Juice, Vinegar
10,000	PH = 3	Grapefruit, Orange Juice
1,000	PH = 4	Acid Rain, Tomato Juice
100	PH = 5	Black Coffee
10	PH = 6	Urine, Saliva
1	PH = 7	"Pure" Water
1/10	PH = 8	Sea Water
1/100	PH = 9	Baking Soda
1/1,000	PH=10	Milk of Magnesia
1/10,000	PH=11	Ammonia Solution
1/100,000	PH=12	Soapy Water
1/1,000,000	PH=13	Bleaches, Oven Cleaner
1/10,000,000	PH=14	Liquid Drain Cleaner

Two Engineers

Two engineers, one a mechanical engineer and the other an electrical engineer, were sitting in a watering hole discussing the complexity of the human body. During the course of the conversation, they both agreed that God must have been an engineer to design the human body. The disagreement came when they argued which type of engineer he was.

"He has to be a mechanical engineer. When you consider the complexity of all the joints and how they work together. In addition, the arteries and veins are just like pipes. The heart is like a pump. He definitely is a mechanical engineer." The mechanical engineer said.

"No way! He is an electrical engineer. Think about all the nerves. They are like wires and sensors. And the brain, it's like a computer. He is definitely an electrical engineer," the electrical engineer proclaimed. The argument became heated when the boiler guy sitting next to them said, "You are both wrong"

They both turned to look at him as he continued, "God must have been a civil engineer," he said slowly

"No way. Why do you think that?" they both shouted

"Think about it. Only a civil engineer would pipe hazardous waste through a recreational area? It has to be a civil engineer" The engineers agreed.

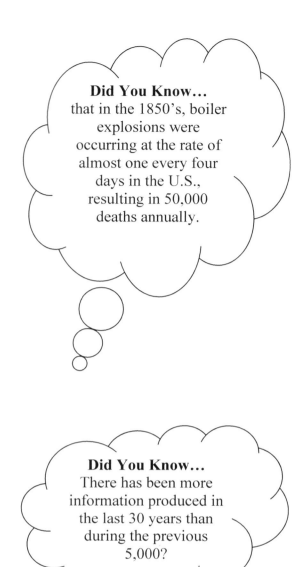

Did You Know…
that in the 1850's, boiler explosions were occurring at the rate of almost one every four days in the U.S., resulting in 50,000 deaths annually.

Did You Know…
There has been more information produced in the last 30 years than during the previous 5,000?

Chapter 3
Boiler Room Walk Through Hydronic Systems

Relief or Safety Valve

The relief valve is probably the most ignored safety control on a boiler. How does a relief valve work? The best analogy that I have ever heard was this: Try to imagine consuming a meal consisting of hot wings, baked beans, beer and carbonated beverages. Your stomach begins to feel pressure from the gas forming

Relief Valve

inside and suddenly it is released, resulting in flatulence. That is what the relief or safety valve does. When pressure forms inside the boiler above the set point, the relief opens and releases the pressure until it drops below the setting of the valve. The system pressure should be at least 10% or five pounds below the setting of the relief valve.

If you did a poll of 100 maintenance people and asked when was the last time that the relief valve was checked, I would wager that 90% of them would say never. I never understood that because the relief valve is the last line of defense before a catastrophe. I would recommend that the building

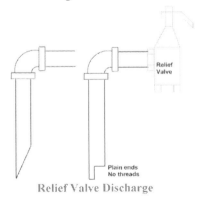

Relief Valve Discharge

maintenance person or custodian check the relief according to the insurance company requirements. This is usually monthly. The relief valve should be mounted in a vertical position. It should be connected to the boiler in as close as a

nipple as you can. If venting above the roof, the discharge piping should be 7 feet above the roof and cut at an angle. Remember, the discharge piping from the relief valve cannot be reduced in size. This reduces the relieving capacity of the relief valve, causing a dangerous condition. The discharge from a hydronic boiler should be vented close to ground and cut on an angle or left unthreaded so that no one will install a pipe cap. The relief valve piping should be supported. The relief valve is not

designed to carry the weight of the discharge piping. Be careful of the relief valve piping. I had a client that piped the relief valve discharge piping to the ground using a copper tube. The tube was tight to the ground. When the relief valve opened, the tubing expanded and broke the relief valve. The boiler room flooded. Conbraco recommends using schedule 40 black iron pipe for the discharge piping on their relief valves. The following drawing is one that my friend Mr. Chuck Ray uses on his boiler projects when piping the safety valve discharge piping with

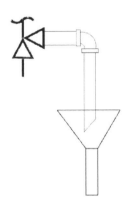

copper. It eliminates the possibility of the copper expanding and cracking the relief valve.

Relief Valve Leaking

A leaking relief valve could be a future service call. The cause of the leak should be investigated. On a hydronic boiler, a weeping relief valve could indicate the following:

- Plugged piping to the expansion tank
- Flooded expansion tank
- Defective water feeder

- Excessive boiler pressure
- Undersized expansion tanks
- Defective bladder in expansion tank

Relief Valve Sizing
The relieving capacity of the relief valve shall be greater than the gross output of the boiler. For example, if the boiler is 1,000,000 Btuh and the gross output is 800,000 Btuh. The relieving capacity must be greater than 800,000 Btuh.

Hydronic Piping Sizes
Pipe sizing is important in a hydronic system. Under sizing the piping is far more detrimental to the system than over sizing the piping. Undersized piping could lead to
- Boiler short cycling
- Reduced seasonal efficiency
- Comfort complaints
- Increased heating costs
- Noisy operation

Oversized piping could lead to increased installation costs.

Normal Water Temperature
Most non-condensing hydronic systems were designed to supply 180^0 F at the outdoor design temperature with a temperature drop or ΔT(Delta T) of 20^0 degrees F. On conventional, non-condensing boilers, operation below 140^0 degrees F on the return or 160^0 degrees F on the supply could result in the condensation of flue gases. Flue gas condensation will destroy the boiler and flue.

The boilers that are designed to condense will tolerate much lower temperatures than a traditional boiler. Condensing boilers do not actually condense until the water temperature is below 140^0. If a condensing boiler is operated above 140^0, it has efficiencies in the mid to upper 80% range.

Circulator
According to industry averages, based mounted pumps will last 20 years and pipe mounted pumps will last 10 years.

Circulators should be replaced when the boilers are replaced. In this way, you will be assured that the circulators will be properly sized for the new heating system. It is also a good idea that you have some sort of redundancy with the pumps in case one pump fails. On large systems, variable frequency drives should be considered since the pump operates continuously during the heating season. For example, a 2 horsepower pump will cost the building owner about $6.46 per day to operate. This would be based upon a kwh cost of $0.1346. If you estimate a heating season of October 1 to March 31, the owner would pay $1,169.26 for the electrical consumption of the pump.

Which is it?
I visited a building where the owner was considering a boiler replacement. They wanted a boiler with the same capacity as the existing one. The boiler had a gross output of 1,200,000 Btuh. During my walk through, I noticed that the circulator for the building was 60 GPM. I decided to investigate further. A 60 GPM pump will deliver roughly 600,000 Btuh, based upon a 20-degree Delta T. The 60 gpm pump was half the flow that the boiler required. This could explain why the boiler was in need of replacement. There was insufficient flow for the boiler. This caused the boiler to experience thermal shock. The boiler temperature rise was 40^0F instead of the 20^0F rise that the manufacturer recommended. The question we had to answer was whether the boiler was oversized or the pump was undersized.

Boiler Flow
Check to make sure that there is adequate flow when the boiler fires. A rule of thumb is that the boiler should have at least three GPM per boiler horsepower. This should be verified with the boiler manufacturer.

System Pressure
One pound of pressure will lift water 2.3 feet. If the highest radiator is 100 feet high above the boiler room, you will need to multiply the height of the highest riser by 0.43. In addition, you will

need an additional 4 pounds of pressure as a safety factor, according to Bell & Gossett. If our highest radiator is 100 feet high, our building will require the following pressure (100 feet x 0.43) = 43 Psi + 4 Psi = 47 pounds of pressure. A quick rule of thumb to see how much pressure we need for a tall building is to divide the elevation of the tallest radiator by 2.

Another aspect of static pressure is the pressure that will be on the boiler if it is located in the basement of the building. In our above example, we stated that the building required 47 pounds of pressure. If that is the case, some cast iron boilers cannot tolerate that much pressure. You may be forced to look at a different type of boiler for this application. That is why many tall buildings will have a boiler room on the penthouse. This limits the static pressure on the boiler. On very tall buildings, you may have several boiler rooms at different floors in the building. Our area was tradition-ally a cast iron

Expansion Tank

market and some engineers used to design a low-pressure steam boiler that would heat the water via a steam to water heat exchanger. This was to avoid the static pressure on the boiler. If you see this type of application, you could save the client money by converting the system to hydronic. *The water feeder should be replaced when the boilers are. This will avoid a potential service call.*

Expansion or Compression Tank
Every hydronic heating system requires either a compression or an expansion tank. Water will expand at 3% in volume when heated from 32^0F to 180^0F. Without a means for handling that expansion, the relief valves will open, spilling water. Water containing glycol will expand at an even greater rate.

Many expansion tanks have a sight glass that shows the level inside the tank. On a cold system, the level is usually about 1" above the bottom of the sight glass. When

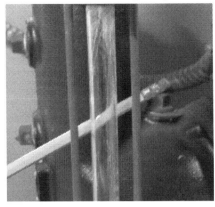

the system water is warm, the level is about ½ to ¾ up the sight glass. If unsure about whether the tank is empty or full, hold a pencil behind sight glass. If the tank is empty, the pencil will appear normal. If the tank is full, the pencil will appear to be broken behind the gauge glass. The picture above shows no water. This technique can also be used when looking at the gauge glass on a steam boiler. A flooded expansion tank could be caused by a couple reasons:

- Fitting leaks above the water level in the tank
- Excessive pressure
- Undersized tank
- Hole in tank
- Defective diaphragm in tank

When using a bladder- type expansion tank, check the air pressure inside the diaphragm using a tire pressure gauge. The air pressure should be tested while there is no system pressure on the tank. If the pressure is too low, nitrogen or oil less air should be used to fill the diaphragm. A sign of a flooded expansion tank will be to watch the hydronic boiler pressure gauge as the boiler fires. If the pressure builds as the boiler fires, it could indicate a flooded expansion tank or plugged piping to the expansion tank.

Suggested Makeup Water Piping Arrangement
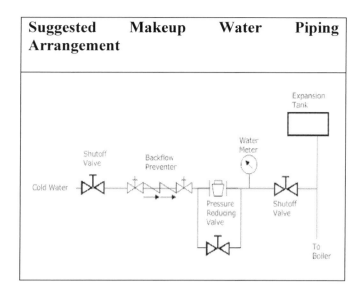

Expansion Tank Size

Expansion tank sizing is discussed later in the book. A rule of thumb for sizing an expansion tank is:

One gallon for each 23 square feet of radiation or One gallon for each 3,500 Btu of radiation.

Reusing The Expansion Tank

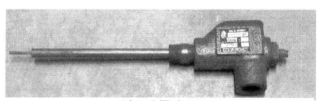

Airtrol Fitting

If you are going to reuse the existing expansion tank, you should inspect the Airtrol fitting. This fitting is in the bottom of the expansion tank. It allows you to drain the water from a flooded tank and also add air to the top of the tank. When reusing the existing expansion tank, remove and reinstall the fittings on the sight glass. They sometimes allow air to escape from the tank.

ASME Tank

When replacing an expansion tank, many commercial buildings require that the expansion tank be ASME rated. ASME is the American Society of Mechanical Engineers. The tanks will be more expensive than a standard tank.

Boiler Control Options
Boiler Design Temperature

Most hydronic systems were designed to provide 180^0 F water at the "Outdoor Design Temperature". The "Outdoor Design Temperature" was originally the 1% design temperature. This meant that in an average winter, the outdoor temperature would be colder than the design temperature 1% of the time. Now, the industry has adopted the 2 1/2% temperature. This means that in an average winter, the temperature will be at or below the design temperature 2 1/2% of the time. For example, if we have 4,000 heating hours during a typical heating season and our design temperature is 0^0F, 120 hours will be below 0^0F. The outdoor design temperature is different in almost every locale. For instance, the heating design temperature for Fairbanks, Alaska is -47^0F while the design temperature for Jacksonville, FL is 32^0F.

Velocity

The velocity in a hydronic system should range from 1 ½ to 4 ½ feet per second in occupied areas. If the flow is excessive, the system will be noisy. The velocity can be slightly higher in unoccupied areas. Flows greater than 6 feet per second can erode copper. In case you were wondering, One foot per second = 0.6818 Miles per hour. 1.5 FPS=1.022 MPH, 4.5 FPS = 3.06 MPH.

To find Fluid Velocity

$$Feet\ per\ Second\ \frac{408\ x\ GPM}{(Pipe\ Diameter\ Inches)^2}$$

Reset Control

Assuming that our original heating system was designed for a heat loss of 1,000,000 Btuh at the design outdoor temperature of 0^0 F with a hot water supply temperature of 180^0 F. The system was most likely also designed for a 20-degree Delta T or temperature drop. That means that the return water temperature will be twenty degrees lower than the supply temperature. Our design was also based on a certain indoor temperature. In this example, we will use 72^0 F. The greater the temperature difference between the outside air temperature and the inside temperature, the greater or faster the building loses its heat to the outside.

As the outside temperature rises, the difference between the inside temperature and the outside temperature decreases. Since the temperature difference is less, our heat loss rate declines. If we used 180^0 F to successfully heat our building at 0^0 F, it makes sense that we could heat our building at 30^0F outside temperature with a lower water temperature. A reset control will assume that as well. A reset control will reduce the system temperature as the outside temperature rises. A common reset ratio in the industry was one to one. That means that the system water temperature drops one degree for every degree that the outside air rises above the design temperature. One of the problems with this reset schedule was that the boiler flue gases would condense if the water temperature dropped to below 140^0 F. In reality, we could only reset the water by 20^0 F without fear of condensation. They were installed on many buildings. Verify with the new boiler manufacturer how low the reset could be set.

Typical One to One Reset Schedule	
Outside Temperature	Supply Temperature F
0^0 Design temperature	180^0 F
20^0	160^0 F*
40^0	140^0 F*
60^0	120^0 F*

*Standard boiler could be condensing at this temperature.

Three Way Valves Three way valves were sometimes installed to help reduce fuel costs. In theory, it was a good idea. In reality, it does not work as well. The boiler would be maintained at the boiler manufacturer's recommended setting. This was typically 160- 180^{0}F. The heating loop to the building would be reset by bypassing the boiler. If the loop required heat, the valve would open and allow water to flow through the boiler. The drawback to this design was that the boiler could not handle the wide temperature span. For

example, the boiler would be set at 180^0 F to avoid condensation. In some instances, the loop may drop as low as 120^0 F. The boiler would see a 60^0 F temperature difference and this could cause rapid expansion and contraction of the boiler. This is called Thermal Shock. It will destroy the boiler. Most boilers are designed for a 20^0F temperature rise.

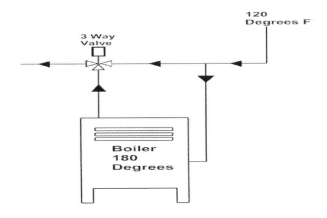

Pressure Temperature Gauges Hydronic Systems
The indicator type of gauge on hydronic boilers is called a Tridicator or a PTA gauge. PTA stands for Pressure, Temperature, Altitude. In the earlier section, we saw that 1 pound of pressure would raise water 2.31 feet. You can look  at this gauge and see the following, height of the water, system static pressure and temperature of water.

Copper Pipe Maximum Hydronic Flow Rates
Based on 20 degree F Delta T

Pipe Size	Maximum Flow GPM	Btuh
½"	1 1/2	15,000
¾"	4	40,000
1"	8	80,000
1 ¼"	14	140,000
1 ½"	22	220,000
2"	45	450,000
2 ½"	85	850,000
3"	130	1,300,000

Hydronic Pipe Sizes
PEX Piping Maximum Hydronic Flow Rates

Pipe Size	3/8"	½"	5/8"	¾"	1"
Max GPM	1.2	2	4	6	9.5
BTUH	12,000	20,000	40,000	60,000	95,000

Steel Pipe Maximum Hydronic Flow Rates
Based on 20 degree F Delta T

Pipe Size	Maximum Flow GPM	Btuh
½"	2	15,000
¾"	4	40,000
1"	8	80,000
1 ¼"	16	140,000
1 ½"	25	220,000
2"	50	450,000
2 ½"	80	850,000
3"	140	1,300,000
4"	300	3,000,000
5'	550	5,500,000
6"	850	8,500,000
8	1,800	18,000,000
10"	3,200	32,000,000
12"	5,000	50,000,000

Chapter 4
Boiler Room Walk Through Steam Systems

Safety Valve

The relief valve should always be mounted in a vertical position. The rating of the relief valve should always be higher than the rated output of the boiler. On steam systems rated at over 500,000 Btuh, the relief valve

Drip Pan Ell

discharge piping should be vented outside. This is often overlooked and could be a very expensive cost that you would have to incur, if missed in your quote.

Steam boiler relief valves use a drip pan elbow on the discharge of the boiler safety valve. The drip pan ell is used to collect condensate that might accumulate in the discharge piping and direct it away from the relief valve. It also limits the stress on the relief valve. If venting above the roof, the discharge piping should be 7 feet above the roof and cut at an angle. Remember, the discharge piping from the relief valve cannot be reduced in size. This reduces the relieving capacity of the relief valve, causing a dangerous condition.

Steam Pipe Fittings

Cast Iron fittings should be used for steam fittings. Avoid using malleable fittings.

Pressure Gauges

Steam pressure gauges and pressure controls are attached to the boiler with a siphon or "pigtail". The siphon or pigtail is a piece of pipe that is bent in a 360-degree angle. The siphon is designed to protect the control or gauge from the live steam. On initial startup, the steam enters the siphon and then condenses. The water is trapped inside the siphon and this water seal will protect the control from the damage.

Steam Boiler Pressure Gauge	
Notice the pigtail or siphon in the piping. This will create a water seal to limit the live steam from ruining the gauge. The loop should be turned so that the loop is perpendicular to the front of the gauge or control.	 Siphon or Pigtail

Steam Boiler Controls

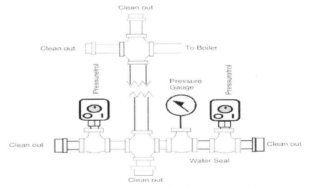

Most low-pressure steam boilers should have two pressure controls; one is the operating control and one is the limit control. The operating control is set for the steam pressure that your building needs, typically 2# of pressure or less. The limit control is set higher than the operating control. The limit control should have a manual reset. This means that the control has a button that has to be pushed to make the boiler operate again. Why would we want this control be a manual reset control? Let us assume that the operating control is set at 2 pounds and the limit control is set at 10 pounds steam pressure. If the boiler pressure raises high enough to trip the limit control(10 PSI), what is that telling you about the operating control? It means that the operating control is not working and may need to be replaced. If the limit control is an automatic reset rather than a manual reset, the boiler could operate on the higher pressure. How long do you think it would take before the higher pressure was discovered? The manual

Union on Siphon

reset forces the owner to look at the boiler in the event of a "No Heat" call. Be sure that you use a siphon or pigtail between the boiler and the pressure control. If you are so inclined, you could make your own siphon from normal pipe fittings.

See the diagram on the previous page. You will see that there are pipe caps on the siphon to allow cleaning of the fittings.

When using siphons, use two different pipe openings for the controls. This reduces the possibility of a plugged pipe disabling both the operating and limit control. If you see the picture above, a plugged pigtail could disable both controls.

Another item that will save the future service technician from cursing you out is to include a union in the pipe from the siphon to the control. See picture above, it will allow the technician to replace the control without having to disassemble

the entire piping from the boiler to the control. The loop on the pigtail should be perpendicular to the front of the control. When steam is applied to the pigtail, it wants to straighten itself. If the loop is sideways, it could affect the pressure setting of the control.

Pigtail is installed wrong here. It could affect pressuretrol settings. See arrows. The far right control is piped correct.

Near Boiler Piping

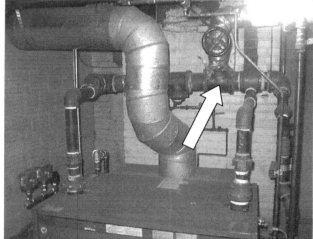

To have a trouble free steam system, the near boiler piping is crucial. Document or photograph the piping. The near boiler piping in the picture is incorrect. The steam takeoff to the building is piped between the two risers coming from the boiler. This leads to wet steam and increased operating costs. The diagrams below show different ways to pipe a cast iron boiler. On Drawing A, you will notice some extra piping called a "swing joint". This is required on cast iron boilers due to the different rates of expansion of the two metals. The piping would expand at a different rate than the boiler and could result in a damaged boiler. On Drawing B, the swing joint is not installed and this could lead to cracked sections.

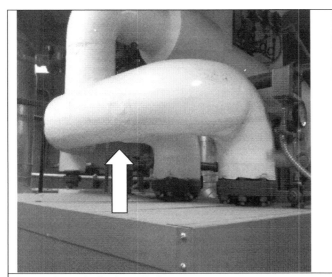

Drawing A Swing Joint

**Drawing B
No Swing Joint**

Steam Cast Iron Boiler Header Piping
Near Steam Boiler Piping

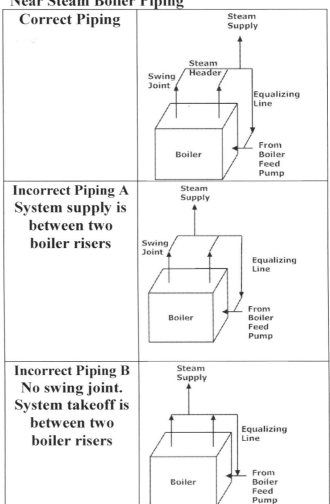

Correct Piping	
Incorrect Piping A **System supply is between two boiler risers**	
Incorrect Piping B **No swing joint. System takeoff is between two boiler risers**	

Near Boiler Pipe Size

Excessive steam velocity should be avoided in a low-pressure steam system. Excessive steam velocity will cause water to be entrained in the steam. This is called "Wet Steam." Wet Steam causes a variety of problems.

Reduced Efficiency – Excessive water in the steam system will cool the steam, causing a drop in system efficiency.

Increased Maintenance – When the wet steam enters the system, the water causes increased maintenance by plugging the steam valve and trap orifices.

Increased Water Treatment Costs – When water from the boiler is carried into the system, it carries with it some of the water treatment chemicals. This causes the treatment chemicals to be consumed and will require replacement.

Check Valve

A good idea when installing steam boilers is to install a check valve in the steam supply piping from the boiler. This will reduce the amount of steam that finds its way into the lag boiler. This will also reduce the jacket losses from the lag boiler into the boiler room.

Vacuum in Steam Piping

We had a project where the owner installed radiator valves on every radiator in the building to save money. When the boiler shut off and the steam began to condense, the steam system developed a deep vacuum in the piping. On some occasions, it would actually pull the condensate from the boiler feed unit and flood the boiler. Once we installed a vacuum breaker, the problems disappeared.

Steam Velocity

One of the leading causes of steam carryover is the near boiler pipe sizing. If the piping is undersized, the steam velocity will be high enough to carry water into the system.

The old timers used to design the near boiler piping for a velocity of 15 fps or feet per second. To keep installation costs down, most newer boilers use a higher velocity and smaller pipes. They use the near boiler piping to dry the steam before it enters the system. I performed a study of 50 of the most popular steam boilers and found that the average velocity in the near boiler piping is 45.77 feet per second or fps. Spirax Sarco recommends a velocity of 40 feet per second.

The following table will assure you that the steam velocity will be 40 fps or less, based on 2# of steam pressure. Each boiler manufacturer requires different velocities. Please check their requirements.

Pipe Sizing to Assure Steam Velocities Below 40 FPS		
Pipe Size	Lbs / Hr	Btu/ Hr
2"	140	134,400
2 ½"	199	191,200
3"	307	295,600
4"	530	509,200
5"	833	799,600
6"	1,204	1,155,600
8"	2,131	2,046,000
10"	3,397	3,262,000
12"	4,783	4,592,000

For example, a boiler rated for 750,000 Btuh output, would require a 5" pipe.

To see the difference between the two designs, here is the capacities if you were designing a steam system with a velocity of 15 feet per second.

Pipe Sizing to Assure Steam Velocities Below 15 FPS		
Pipe Size	Lbs / Hr	Btu/ Hr
2"	52.5	50,400
2 ½"	74.69	71,700
3"	115.47	110,850
4"	198.91	190,950
5"	312.34	299,850
6"	451.41	433,350
8"	799.22	767,250
10"	1,274.22	1,223,250
12"	1,793.75	1,722,000

For example, a boiler rated for 750,000 Btuh output, would require an 8" pipe.

Steam Header Pipe Velocity

The above study also showed that the boiler headers had an average velocity of 55 feet per second. The highest velocity was 65 fps and the slowest was 47 fps. The following is a design guide using 50 feet per second.

Pipe Sizing to Assure Steam Velocities Below 50 FPS		
Pipe Size	Lbs / Hr	Btu/ Hr
2"	175.00	168,000
2 ½"	248.96	239,000
3"	384.90	369,500
4"	663.02	636,500
5"	1,041.15	999,500
6"	1,504.69	1,444,500
8"	2,664.06	2,557,500
10"	4,247.40	4,077,500
12"	5,979.17	5,740,000

How Fast is Fast?

One foot per second = 0.6818 miles per hour. The following is a chart to help you to compare the steam velocities.

Feet per Second to Miles per Hour				
FPS	MPH		FPS	MPH
15	10.2		50	34.09
20	13.6		55	37.5
25	17		60	40.9
30	20.4		65	44.31
35	23.86		70	47.72
40	27.27		75	51.13
45	30.68		80	54.54

Trap the Steam Header

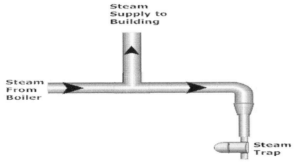

As steam leaves the boiler and travels throughout the piping, it loses some of its energy. When this happens, some of the steam will condense. To assure dry steam, the main steam supply pipe should trapped at the end of the main. When trapping the header, pipe the header so that it is the full pipe size until you get past the steam supply takeoff to the building and the elbow back to the boiler feed tank. See drawing above. You

also want to remember the proper static head above the trap.

Some near boiler piping rules of thumb

- Steam supply piping to building should always come from the top of the steam piping.
- The steam header should be pitched back toward the equalizing line.

A Couple Thoughts on Steam Traps

When trapping the steam header, there are a couple items to remember. The first is that the feed to the trap must be a full size diameter tee that will feed the trap. If you welded a 3/4" tapping into the bottom of a 8" header, most of the water will zoom past the tapping. It would be like driving down the road at 30 miles per hour and reaching out the door to pick up a quarter on the road.

Another item to remember is that the pipe to feeding the steam trap should be at least 15" below the steam header. This will add a static head to the trap to allow it to work better. A 15" static head will provide a 1/2 psi pressure differential across the trap when it drains into a gravity return system that is vented to atmosphere. This is required to drain the trap when the boiler is off and the steam pressure is at 0 psi.

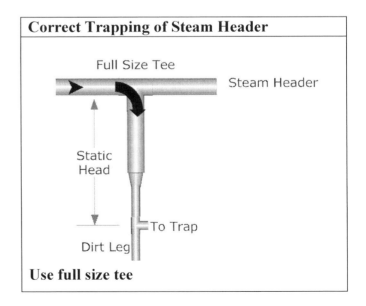

Correct Trapping of Steam Header

Use full size tee

Incorrect Trapping of Steam Header

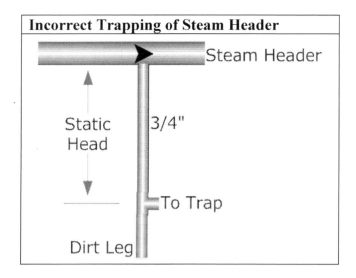

Connecting Boiler to Building Piping

When connecting the new steam boiler to the piping, the steam piping should connect to the top of the steam header. If the steam piping is connected to the bottom of the header, the condensate in the piping will drop back into the boiler and could cause flooding. If you cannot pipe the steam into the top of the steam piping, the riser should be trapped so the condensate does not drip back to the boiler.

Cast Iron Boiler Piping

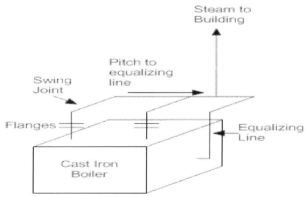

A very accomplished installer, Paul Lancaster, gave me this bit of advice for when he installs cast iron sectional boilers. He installs flanges in the steam supply piping to allow for easier access to the boiler sections. Instead of having to cut and re-weld the steam piping, the flanges could be unbolted for service.

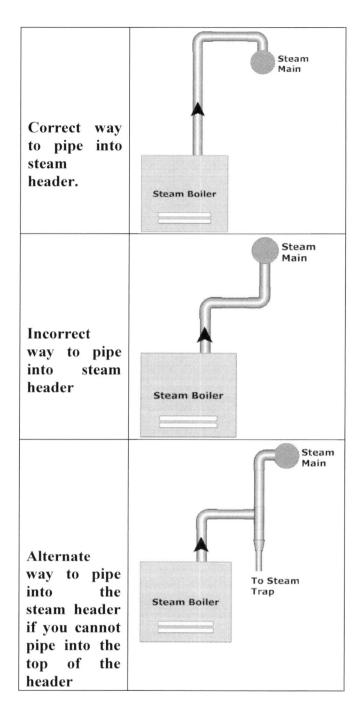

Correct way to pipe into steam header.

Incorrect way to pipe into steam header

Alternate way to pipe into the steam header if you cannot pipe into the top of the header

Condensate Pipe Size

Most steam systems rely on gravity to return the condensate from the system to the condensate tank. Some old steam systems used a vacuum pump to pull the condensate back to the boiler. An advantage to a vacuum return was that the condensate piping could be undersized since the pump would pull the condensate back to the boiler. In addition, the vacuum pump could overcome any of the minor dips in the piping.

These vacuum type systems were costly to maintain and most were abandoned in favor of a standard condensate tank, which rely on gravity to return the condensate. The drawback to converting a vacuum system to a gravity return is that the condensate piping could be undersized. This leads to a slow return of the condensate and possible flooding of the condensate tank or boiler. An example of this would be if we size a boiler feed tank for 15 minutes storage. That means that we think it will take 15 minutes for the condensate to return from the system once the boiler is started. If it takes longer than that, the tank will run out of water. The internal float will open and feed fresh, untreated water into the tank. Eventually, the condensate will return. When it does, the tank will flood and the water will spill out of the vent. This will waste the heat as well as the chemicals. If you are unsure as to whether the system used to be a vacuum system, a slightly oversized boiler feed tank will avoid these problems.

Condensate Pipe Note

Most of the old steam systems used schedule 80 pipe on the common condensate return piping. Schedule 80 piping is about 50% thicker than the standard schedule 40 pipe. The old designers knew that carbonic acids would form in the condensate lines as a result of the carbon dioxide that mixes with the water in the return piping. If your replacement consists of replacing return piping, remember that you should consider schedule 80 pipe.

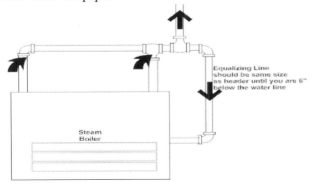

Equalizing Line

The equalizing line equalizes the pressure inside the boiler from the steam side to the condensate side. This leads to a stable water level. The equalizing pipe should be the same diameter until it is below the normal boiler water line. The equalizing line should be at least 2 ½" in diameter. If the boiler is rated for over 2,500,000 Btuh, the equalizing line should be at least 4".

Boiler Feed or Condensate Tank

When replacing a steam boiler, most boiler manufacturers recommend replacing the condensate tank with a boiler feed unit. On the older steam systems, the condensate tank operated independently of the boiler. The condensate tank had an internal float. Whenever the water level rose in the tank, the float would energize the pump. The pump would deliver the condensate inside the tank to the boiler, regardless of whether the boiler needed water. If the heating system required makeup water, it was fed through a water feeder control directly to the boiler. It is not good practice to feed cold city water into a warm boiler.

The new boilers subscribe to a different type of philosophy regarding condensate. A boiler feed system is used instead of a condensate tank. The difference between the two is that the boiler feed unit is usually much larger than a condensate tank. It is typically about five times larger than a condensate tank. The cold makeup water is fed into the boiler feed tank where it will be preheated and chemically treated before being introduced into the boiler. This reduces the chances of thermal shock to the boiler. Another advantage is that the city water, when mixed with the hot condensate, will release some of the oxygen into the boiler feed tank. Since the water level is critical on the new boilers, they incorporate a combination pump control and low water cutoff. When the boiler water level drops to a certain level, a set of contacts inside the pump control will energize the boiler feed pump, filling the boiler.

Duplex Condensate Tank
Courtesy: Sterling Steam Control Products

Float

Duplex Boiler Feed Tank with Spare Pump

In case you were wondering, a duplex boiler feed means that the unit has two pumps.

Steam Boiler Make Up Water Piping
On the makeup piping for the steam boilers, a pressure-reducing valve should be installed in the piping. The internal makeup valve inside the boiler feed tank will sometimes require lower pressure than the city water pressure to the facility. If the water pressure is too high, it will overcome the internal float valve and flood the boiler feed tank.

Steam from Boiler Feed Vent
If steam is coming from the boiler feed vent pipe, it could indicate a couple problems. The first is that the steam traps could be leaking through. Leaking steam traps could allow steam to escape and makeup water would have to be introduced. Leaking steam traps could also cause some areas to have comfort complaints. The second problem could be excessive steam pressure. Float and thermostatic steam traps are designed to open once the temperature drops 20 degrees F. If the boiler is set at 7# steam pressure, the temperature of the steam is 232 degrees F. That means that the condensate will be discharged into the pipe at 212 degrees F. As you are aware, that is the boiling temperature of water so the condensate could flash to steam. When water turns to steam, it expands 1,600 times its volume. The flow of condensate back to the boiler slows and the pipe is filled with steam. This boiler feed unit runs out of water and feeds fresh, untreated water into the system. It also increases the temperature inside the condensate tank. Elevated condensate temperatures could cause damage to the boiler feed pump. If the water is too warm, the condensate could flash to steam inside the pump vortex and destroy it. This is called cavitation and it sounds like the pump has marbles inside it. The steam will destroy the pump's impeller and reduce pump capacity. In many condensate pumps, the impeller is made of a softer metal like bronze. It could also attack and destroy the pump's mechanical seal. In severe cases, the boiler may go off on low water cutoff.

Back Pressure on Steam Trap Capacity.
Leaking steam traps will fill the condensate pipe with steam, reducing the pipe's capacity. The steam in the condensate pipe is called "Back Pressure". If the system has steam in the condensate piping, the trap relieving capacity is diminished. This lowers the efficiency of the system. The following chart details the effect of back pressure on the operation of steam traps.

Effect of Back Pressure on Steam Trap Capacity	
Back Pressure	**Loss of Capacity**
25%	6%
50%	20%
75%	38%
Based upon 5# steam pressure	

Air in Steam Lines

Air in the steam pipes will reduce the efficiency as well as the temperature of the steam. Properly working air vents could help to eliminate the air. Please see the effects of air in a steam system listed below.

Air in Steam Lines				
Steam Pressure	Pure Steam No Air	5% Air	10% Air	15% Air
2 #	219^0 F	216^0 F	213^0 F	210^0 F
5 #	227^0 F	225^0 F	222^0 F	219^0 F
10 #	239^0 F	237^0 F	233^0 F	230^0 F

Bouncing Water Levels

Improper chemical treatment, oil, excessive demand, excessive solids, or a vacuum could cause bouncing water levels inside a boiler. Bouncing water levels could cause the boilers to trip the low water cutoff and could cause carryover.

If the boilers are chemically treated, check the chemical levels as directed by either the boiler manufacturer or chemical treatment company. The water treatment company can tell you if the solids are elevated in the boiler.

Oil will accumulate in the boilers if they were improperly skimmed or cleaned at startup or if some of the system piping was replaced. When pipe is threaded, the installer uses cutting oil for the pipe threader. This oil can accumulate inside the boiler after several days. Oil is sometimes difficult to remove from a boiler as it adheres to the walls inside the boiler when drained. It may take several attempts to flush the boiler to rid it of the oil. Check with the boiler manufacturer when using a chemical to clean the boiler to verify that it is safe to use on their equipment.

Another cause of bouncing water levels could be zone valves on the steam supply lines. If the zone valves are not installed properly, a vacuum could occur in the piping and could literally suck the water from the boiler when the valve opens. Even if the valves are installed correctly and there is no vacuum, a quick opening valve opening could affect the water levels. If the piping is cool and the valve opens, the steam will rush to the cool piping and cause a surge in the boiler. This could affect the steam velocity and pull the water from the boiler.

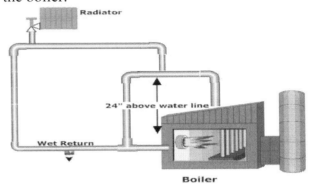

Without Hartford Loop

Hartford Loop

When coal was used as a fuel source in the old boilers, control was rather difficult. The old coal boilers could not be shut off like the new gas or oil boilers. If there was a leak in the wet return piping, the boiler could lose all its water and dry fire, which is very dangerous.

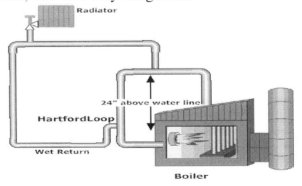

With Hartford Loop

See drawing above, Without Hartford Loop. The Hartford Loop was developed to prevent the water from draining the boiler. With a Hartford loop, a leak on the wet return would not drain the boiler completely of water. It would drop to the elevation of the Hartford Loop fitting. On most commercial boilers with modern safety controls, they are no longer needed. They are still installed on steam boilers without a feed water pump or condensate tank.

Checking the Gauge Glass

Sometimes, it is difficult to tell by looking at the gauge glass if the boiler is flooded or empty. If you hold a pencil behind the gauge glass, it will give you a clue. If the gauge glass is empty, the pencil will look normal. If the gauge glass is filled with water, the pencil will look broken behind it.

Empty Gauge Glass

Two Feet to Drier Steam

Newer steam boilers use the near boiler steam piping as a drying chamber for the steam. The industry has adopted 24" as the height that the steam header should be above the water line on the boiler. This helps dry the steam to keep the heating costs and maintenance low.

Double Trapping

When you see a trap on the condensate pipe at the entrance of the condensate or boiler feed tank, be very careful. You might be inheriting a problem job. The problems most likely started when one or two traps started to leak steam. The boiler feed vent would then start spewing steam into the boiler room. Someone then decided to install a "master" trap at the inlet to the condensate or boiler feed tank instead of repairing the traps. That is when the system complaints start. The person then tried raising the steam pressure to overcome the double trapping. The clients fuel bills start to soar and the comfort

complaints keep coming in. If you see a trap on the inlet to the boiler feed, you need to inform the client that there are system problems that should be investigated. There are rebuild kits for traps that are relatively inexpensive. This could be an opportunity to get more work from the client.

Double Trapping

Lag Boiler Flooding

A common problem when the heating plant has two or more boilers is that the lag boilers can flood. When this happens, there could be water hammer and in some instances, the boiler will not fire. This occurs because the boilers are connected to a common steam header. The steam will enter the boilers that are not firing and will condense. It will eventually cause the water level inside the idle boiler to rise. If you remember from an earlier chapter, it takes one pound of pressure to cause the water to raise 2.3 feet. Conversely, if we have six feet of water inside the boiler above the pressure control, we will have 2.6 pounds of static pressure on the boiler pressure control. If we are operating the boilers at 2 pounds of pressure, that boiler will never fire until we drain the boiler.

The American Society of Mechanical Engineers or ASME recognized this as a problem and offer the following solution. It is called a "High Level Spill". The High Level Spill is a steam trap that is installed about 1" above the normal water level. If the water level inside the boiler rises to the elevation of the steam trap, the steam trap will open and drain the water from the boiler.

The discharge of the trap could be piped to the boiler feed unit so that the water is not wasted.

Some water treatment companies do not want the discharge from the high level spill piped to the condensate return system as it appears as carry over when they are doing their tests. If the boiler feed tank is far away, a condensate tank could serve as a transfer pump. The common discharge of the high-level spill traps could be piped into a condensate tank that will pump them back to the boiler feed tank.

A High Level Spill Lesson.
We were called to a job that was experiencing boiler flooding. The boilers had high level spill traps installed on each boiler. The discharge of each high level spill trap rose three feet before going into the condensate pipe. In addition, there was no check valve on the discharge of the piping after the trap. A check valve is required if you want to "lift" the condensate. The high level spill trap is designed to drain the lag boiler that is not fired. If the boiler is not fired, there will not be enough pressure to lift the condensate, resulting in a flooded boiler. The installer felt that the steam pressure from the other boiler would lift the condensate. The problem with that is that a one ounce difference in steam pressure between the two boilers would result is a water height difference of 1 3/4". In addition, if both boilers were satisfied and shut off, there was no way to lift the condensate and the boilers will flood. If the flooding is too high, the boilers may not start due to the static head on the pressure control.

Low Water Cutout

Combination Low Water Cutoff Pump Control

Commercial steam boilers will use two low water cutoffs. Please notice on the picture above that the installer used pipe caps instead of pipe plugs on the piping for the low water cutoff. That will make it much easier to disassemble and clean the openings. The first low water cutoff is usually a combination pump control and low water cutoff. The second is the auxiliary low water cutoff which usually has a manual reset feature. As the boiler water level drops, the first set of contacts will energize, starting the boiler feed pump. If the water continues to drop, the second set of contacts will open, cutting power to burner. If the water continues to drop, the contacts on the auxiliary low water cutoff will open, cutting power to the burner. The auxiliary low water cutoff should be a manual reset control that will need to be reset before the power to the burner is restored. The reason for a manual reset on the low water cutoff is to alert you that the primary low water cutoff did not operate correctly. The auxiliary low water cutoff is installed at a lower elevation than the primary low water cutoff. Some low water cutoffs have an audible alarm that will notify the owner of a low water condition.

Be careful connecting the boiler feed pump to the combination pump control/ low water cutoff contacts. The contacts may not be able to handle the amperage of the pump. In most instances, you will need to have the low water cutoff connect to the coil of a relay.

Low Water Cutoff Water Feeder

Auxilliary Low Water Cutoff

Normal Steam Operating Pressure

When heating water, it roughly takes One Btu to raise one pound of water one degree F. That is, until the water temperature reaches 212^0F at atmospheric pressure. The boiler keeps heating that same pound of water trying to make steam. It requires about 970 Btus to get that pound of water to change to steam. This is called Latent Heat. The steam leaves the boiler at about 40 miles per hour looking for something cold. When it finds a cold radiator, it will release that 970 btus and revert back to water or condensate. The real heating of the building occurs when that heat is given up. Most old steam systems were designed to heat the building with less than 2# of steam pressure. If the current boiler is set at a higher pressure than that, investigate why. It could cost the owner increased fuel costs.

Some thoughts about the old steam systems

When the old steam systems used coal as their fuel source, they were designed to run all day long. The boiler tender would keep the fire going inside the boiler. It was rare in the winter that the pipes were allowed to cool. When using a new boiler, the pipes cool in between the firing sequences. This could lead to carry over and bouncing water levels.

Steam Pipe Sizing Charts

Steam Pipe Sizes Steam Main

Pipe Size	Btu/Hr		Pipe Size	Btu/Hr
2"	155,520		4"	912,000
2 ½"	247,680		5"	1,612,800
3"	446,400		6"	2,707,200
3 ½"	643,200		8"	5,347,200

Sizing a Dry Return Piping System

Pipe Size	Btu/Hr
1"	98,800
1 ¼"	208,320
1 ½"	326,400
2"	710,400
2 ½"	1,180,800
3"	2,160,000
4"	4,636,800

Size Steam Main by Connected Radiation

Radiation Sq Feet	Pipe Size Inches
75-125	1 ¼
125-175	1 ½
175-300	2
300-475	2 ½
475 – 700	3
700-1200	4
1200-1975	5
1975 – 2850	6

Tales From The Field

I was on a job site to perform a startup of one of our steam boilers. While watching the boiler run, it built up steam pressure very quickly and shut off. I found it odd that it was so quick. I touched the steam pipe going to the building and it was cold. The school maintenance man was there and asked, "What's wrong with your boiler?"

I said "There is nothing wrong with the boiler."

"It ain't heating tho. It has to be your boiler. We should have bought the cast iron boilers and not yours." He replied. The maintenance man liked cast iron boilers and had been on record telling everyone that would listen that cast iron boilers were the best. It did not hurt that his brother was the representative for a company that just so happened to sell cast iron boilers.

The contractor was on the job and I asked him if it could be the gate valve. We had pressure up to the steam gate valve and none after. "That is a brand new valve. It has to be your boiler."

I convinced the contractor to help me remove the six inch gate valve. (Those are really heavy, by the way). Inside the valve, we found a piece of plastic that was installed at the factory for shipping. We removed the piece of plastic and reinserted the valve in place. The boiler worked great.

Did You Know…
Did you know that a drop that occurs 30 times per minute will lose 1,041 gallons per year

Chapter 5
Questions to Ask

There is a great deal of work in the design and pricing of a new heating system. Just to give you a warning; If you are lucky enough to win this job, everything that happens in the building will be your fault. I have been in this business long enough to be blamed for everything from the clock in the boiler room stopped working after our boilers were installed to the garage door stopped working right after the boilers were installed.

Questions

1 At what outside temperature does the second boiler start?

Can one boiler heat the building? This will help you to verify your heat loss or boiler sizing calculations. For example, if the existing system has two boilers each rated for 1,000,000 Btuh and the custodian tells you that they have never fired the "lag" boiler; you can **almost** assume that 1,000,000 Btuh will safely heat the building.

2 Are there areas of the building that are hard to heat?

If the person says yes, you should investigate further. On a steam system, it could mean defective valves or traps, unvented areas, air bound radiators or plugged lines.

On hydronic systems, it could mean defective valves; plugged piping or air bound piping. If you are not sure about the problem, you could add a disclaimer in your quote that the problem area will be invoiced extra.

Be careful of broken gate valves. The valve could appear open with a risen stem when the gate broke off and the valve is actually closed.

3 At what temperature do the boilers shut off?

Most buildings can shut off the heat to the building when the outdoor temperature reaches between 55 and 65 degrees F. If there is not a control to automatically shut off the equipment at a certain temperature, this may be an opportunity to provide a quote for one. The control should also shut off the pumps on a hydronic system.

4 Who services the equipment?

A friend of mine installed a very complicated heating system that was designed by an engineer that served on the board of directors of the facility. After two years, the facility called and said that their fuel costs were high and could not understand why. Of course, they blamed my friend, the installer, for the problem. Upon arrival, the installer found that the boilers were running "wild" on the operating controls. The digital control system was disabled. Upon further investigation, my friend found that a Russian immigrant that spoke no English took care of the heating system. He could not understand the controls and manual because he did not speak the language. This was an opportunity to have the client sign a service agreement with his firm.

5 Why are you replacing the boiler?

Some people think that a new heating system will pay for itself in a year or two. That is usually not the case. If the client has unrealistic expectations, this should be discussed prior to the contract being exercised. Typical energy savings are 15-35% of the heating costs with a newer, more efficient boiler. A client I know replaced his cast iron boiler with an atmospheric burner and a standing pilot with the same boiler. He did not see any savings and was upset. If you are replacing the boiler with the same type of boiler, there will be little or no energy savings. Be careful not to guarantee a savings unless a thorough engineering study is performed. If the next winter is more severe than the last one or if the price for heating fuel increased, the client would not save money. We installed new boilers for my church and the business manager called to say that they did not see any savings with the boilers. I did a study of the gas bills and found that they had indeed cut their costs by about 30%. The savings did not show up in their gas bills because their price for natural gas increased 30% in one year.

6 What kind of boilers do you like?

If the client likes cast iron boilers and you are proposing a different type of boiler, this should be discussed. Later in the book, there is a list of advantages and disadvantages on each type of boiler.

7 Is Reliability Important?

If the client has one boiler and is contemplating replacing the boiler with one boiler, you could educate the person about the benefits of two or more boilers.

8 Are Parts Availability a Concern?

Many boiler manufacturers use proprietary parts. This can lead to extended down time and expensive repairs. Clients do not want to be held hostage by the boiler manufacturer. If the equipment that you are proposing has proprietary parts, you may suggest having your client purchase some common repair problems such as ignition electrodes.

9 Does the existing boiler heat the building on the coldest days?

This is something you should know before sizing your new heating system. This will help you with your heat loss calculations.

10 How would you reduce the fuel costs for the facility?

According to a study by the National Association of State Energy Officials, 10-15% savings are possible through operations and maintenance. The other key point of the study is that the staff knows where the savings are. All you have to do is ask.

11 What temperature or pressure are the controls set for?

We sold boilers for a religious facility and designed the system for 180^0 F at the outdoor design temperature. When we checked their usage over the first winter, their fuel costs actually rose. Upon some further investigation, we found that their old boiler operated at 140^0 F throughout the heating season. We were able to reduce the temperature of the new boilers.

As previously discussed, most steam systems were designed to operate at 2 pounds of steam pressure. If the boiler is set higher than that, investigate why.

12 Does Boiler Operate Year Round?

Some facilities operate their boilers year round for reheat coils. This may be an opportunity to consider condensing boilers if the reheat temperature is low or else summer boilers that are smaller and more efficient.

13 How much does it cost to heat the building?

This will give you an idea of the costs per square foot. There are several publications that list average costs per square foot to heat the facility. This will also give you an idea of the possible energy savings. For example, if the heating costs are $50,000.00 per year and your new boiler will save 20% over the existing boiler, the client will reduce his fuel costs by $10,000 per year. If the estimated installed cost is $30,000, the client would have a 3 year payback on the equipment. Another cost to factor in is the increased repair costs that the client would avoid with a new boiler.

14 With an expenditure this high, what does the purchasing cycle look like?

The customer may say something like, "Anything over a certain amount, we need to take three bids" or they may say, "Anything under a certain amount can be done on a negotiated basis." I like to do everything in my power to keep the project in my control. For example, if the purchasing threshold is at $25,000.00 and you anticipate a cost of $40,000, you could break your quote into two options. One option could be for $20,000 for the boiler and the other for $20,000 for the installation.

Another common reply is that the project has to use prevailing wages if over a certain amount of money. Prevailing wages means that the installer has to pay the employees the same rate as the union personnel in that area. This will allow the labor rates to be even for all the bidders on the project. If your company is not a union contractor, you want to make sure that your costs will be at the union scale.

In some buildings, the owner must contact their architect or engineer when replacing a major component like the boiler. The engineer will perform a design and provide specifications and drawings for the project. In this way, all the bidders are bidding "apples to apples"

Have you ever wondered how "Three" became the magic number when getting bids? Why not Two or Four? In any event, this is something that should be known before you invest time and money on a design and proposal.

15 Do you have to take the lowest bid?
If the answer is no, you may be able to work with the client. If the answer is yes, you have to make a tough decision about how or whether to price the job.

* How much will it cost?
Customers will sometimes pressure you to give them a price. My favorite is "Just give me a ball park number." My response to the above question is "Well, PNC Park in Pittsburgh cost $216 million and Yankee Stadium in New York cost $1.5 Billion but I think your heating system will be much less than that. For me to give you a price, I will need to properly design and size the new system. "

Do not give a "ball park" price. It will come back to haunt you. To prepare the client for the final price, consider bracketing the price a little high. For example, if the project will be $20,000 you could say "The boiler replacement will be anywhere from $20,000 to $30,000. When your price is close to $20,000, you look like a hero.
The customer will always remember the higher price. It is at this point that the project will take one of two routes. The first is that the client will encourage you to proceed with your pricing. The second is that you will hear the famous saying, "I need to get three quotes."

What do we do with the information from our walk through?
I would suggest writing a Boiler Room Status Report with pictures of your findings and meet with the potential client. This will do a couple things. It will differentiate your company from the "Low Bidder" that will simply replace the existing boiler(s) without helping the client to fix some long-standing problems. You will become a consultant in the client's eyes. The report is a way of doing that. In addition, it will protect you from possible litigation. Whenever you hand that report to the client, he or she has to do something with the report. They will either send it to a superior or have the extra work done.

Don't Spill your Candy in the Lobby
When I was a new sales representative in the industry after being a service technician for years, I had a potential client that asked me to help him solve a boiler problem. I explained my solution for the problem. The contact innocently asked me to show him a drawing so he could understand what I meant. He took notes about what I suggested. When I was done, he told me that he would let me know. I beamed proudly as I walked out the door, knowing that I had helped my new "client". A few days later, my boss called me into his office. He showed me a request for a quotation from my new "client". It featured my hand drawing and asked me to bid on the project.
"Did you get paid to solve his problem?" he asked
"Uh, no." I replied.
"Hmm" he said and rubbed his chin
"I have a feeling that this will be one of those 'Life's Lessons' that you will never forget." He continued.
"No way, he said he would use our company if I helped him," I protested. We lost the job to a low bidder. The worst part of the lesson was that the "client" called and asked me to clarify something on my design that his installer was confused by. Be careful that the client does not give your report to your competitors. That is where I learned Ray's Rule #5, Some Clients Lie.

Monkey's Paw

When a large ship is docking, the dockhands attach the ship to the dock with huge ropes that will hold it in place. To get the rope to the dockhands, the ship's crew throws a small rope tied in a ball, which is called a Monkey's Paw, to the dockhands. When the dockhands get the smaller rope, they pull it and the smaller rope is attached to the large rope. That is how they get the large rope to the shore. This same approach can be done on the sales call. Instead of asking for the entire job at once, ask for a smaller contract to perform a survey or combustion analysis. This will allow you both to see whether you can work together. If the customer is difficult to work with on the small contract, imagine what he or she will be like on the large contract.

How much experience do you have?

My friend that owned a business was looking to hire a new sales person for his company. One resume went to the top of the potential employees. This individual had twenty years experience in the industry. My friend and the service manager interviewed the potential candidate. After the interview, my friend and the service manager were discussing the interview they had with the potential new hire and the service manager said, "He has 20 years experience. I think we should hire him" My friend said, "He doesn't have 20 years experience. He has one year's experience 20 times. He doesn't seem to learn" We should all try to learn as much as we can about our industry.

Boiler Room Checklist

Boiler Manufacturer	Model/Serial #
Steam/Hydronic?	Input
Fuel	Door Opening Size
Combustion Air Opening	Pipe Insulation?
Operating Temp/Pressure	Gas Pipe Size
Gas Pressure	Breeching Size
Gas train components vented to outside	Backflow preventer
Relief valve piping	Venting problems
Draft controls	Electric available
Exhaust fan	What is stored in boiler room?
Water meter on makeup pipe	Is there ac in room?
Does boiler operate year round?	
Hydronic	
Condition of circulator How many pumps and age	Existing hydronic piping Can you isolate flow through idle boiler?
Expansion tank flooded? Size?	What controls the boiler?
Steam	
Near boiler piping include drawing	Steam boiler pipe size
Condensate pipe size	Equalizing line size
Condensate or Boiler feed unit	
Is relief valve vented to outside if over 500,000 Btuh	Is there steam coming from boiler feed vent?
Gauge glass is it full? Dirty?	Are water levels bouncing?
Is lag boiler flooding	Double Trapping?

Chapter 6
Hydronic Boiler Sizing

Hydronic System

When replacing a hydronic system, perform a heat loss on the building and size your new heating system for that size. A heat loss calculation is not an exact science. Many variables could distort your calculations such as insulation hidden in the walls or the lack of it. This is why a walk through is very important. Balance your heat loss calculation with the actual job site conditions like pipe sizing. If you are able to clock the gas meter, it will give you a true idea of how much fuel is being consumed. A heating system rule of thumb is 20-60 Btuh per square foot with 25-40 Btuh per square foot being the average.

A designer will typically design a heating system for the 97.5% design temperature. This means that you will design a system that can heat the building for 97.5% of a typical winter. For example, in Pittsburgh, PA, where I live, the design temperature is 5 degrees F. That means that about 30 hours or 2 ½% of a typical winter will be at or below that temperature. Usually the colder temperatures occur at night and are not noticed. There is typically a safety factor built into the design to protect the engineer. It is typically 10-20% extra.

Why not choose a larger system? There are three reasons for that. Firstly, industry practices and codes dictate that the designer choose a system that is sized properly. Secondly, many building inspectors want to see a copy of your heat loss calculations and the equipment size before they grant a building permit. Lastly, the equipment and fuel costs will be higher for an oversized heating system.

Controlling Your Hydronic Boiler

A short history of the control of hydronic boilers.
Many years ago, there was a great battle in our land. The battle featured the following participants:

- The Stingy Building Owner
- The Stodgy Boiler Manufacturer
- The Evil Oil Baron
- The Arrogant Wizard

Our tale starts when the Evil Oil Baron raised his prices, citing stockholders needs and CEO entitlements. The Stingy Building Owner cried out in despair. "I cannot pass these costs onto my tenants. Won't someone please help?"

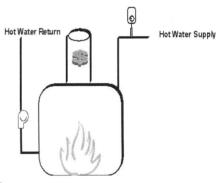

He visited the Stodgy Boiler Manufacturer and asked what he could do. Stodgy Boiler Manufacturer shrugged his shoulders and said that the boiler needed to be at least 160^0 F on the supply and 140^0 F on the return. He rambled on and on about coefficients of expansion and flue gas condensation. No one ever understood what he said.

Then one day, the Stingy Building Owner met the Arrogant Wizard who said, "I can reduce your fuel costs" Stingy embraced the Arrogant Wizard and asked,
"What is it?"
"It's technical. You would not understand." The Arrogant Wizard replied
"Will I lower my fuel costs?" The building owner asked
"Yes, it will" replied the Arrogant Wizard.
"Is it expensive?" the building owner asked.
"Yes, it is very expensive." the Arrogant Wizard replied.
Since Microsoft Excel was not yet invented, the Arrogant Wizard stacked rocks atop each other to show the savings. They agreed and Stingy Building Owner gave the Arrogant Wizard lots of money to help him reduce his fuel costs. The Arrogant Wizard installed a boiler reset control that lowered the temperature of the boiler as the

outside temperature rose. Stingy Building Owner was ecstatic over his "proposed savings"

Entering the new heating season with a notice that the Evil Oil Baron again raised his prices citing a global conflict in Liechtenstein, Stingy Building Owner noticed that his hydronic boiler had problems with leaks and the flue was rusting. It was very expensive to repair. Stingy Building Owner called the Arrogant Wizard and said, "I thought that you said that I would save money."
"First of all, it is not my controls. It is the boiler's fault. Secondly, I said I would reduce your fuel costs. Were your fuel costs lower this year?" the Arrogant Wizard asked.
"Yes, they were $5,000 lower, but the boiler repair costs are $10,000"
"I did as I promised." The Arrogant Wizard replied.
Stingy Building Owner was furious and called Stodgy Boiler Manufacturer and told him what happened.
"I told you so. The minimum entering water has to be at least 140^0 F." Stodgy Boiler Manufacturer said. He went on to explain that hydronic boilers could not have water temperatures below 140 degrees F or the flue gases would condense, ruining the boiler and flue. Since the heating system was designed for a 20-degree delta T and a 180-degree design temperature, the boiler reset could only be reset by 20^0F, to 160^0F on the supply, He suggested to Stingy Building Owner that he call the Arrogant Wizard.
"That Stodgy Boiler Manufacturer is such a prude. Well, if he needs to maintain at least 160^0F in the boiler, I have another solution." The Arrogant Wizard replied with a wink.

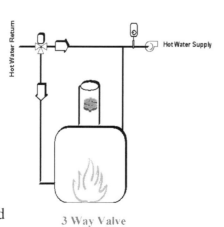

3 Way Valve

"I can help you because I like you. I would suggest a 3 way." The Arrogant Wizard said
"Whoa, there fellow! I'm a God-Fearing, married man and do not indulge in that sort of thing" Stingy Building Owner replied.
"No. What I meant is that we could install a 3-way valve in your system. This will allow the boiler temperature to operate between 160 and 180 degrees to make Stodgy Boiler Manufacturer happy. We can reset the loop temperature down according to the outside air temperature. How does that sound?" the Arrogant Wizard asked.
"Great! How much will this be?" Stingy Building Owner nervously asked.
"Oh this is expensive but think how expensive the fuel costs will be if we do nothing. Let me show you the pile of rocks again." The Arrogant Wizard said. The 3-way valve was installed.

The next heating season led to more expensive repairs including cracked sections and numerous leaks. The Stingy Building Owner realized that he saved $5,000.00 in fuel costs but faced a $11,000.00 repair cost. The Stingy Building Owner was introduced to "Thermal Shock". (*This occurs when there is a wide temperature difference between the supply and return water. It causes rapid expansion and contraction of the boiler, causing leaks.*)

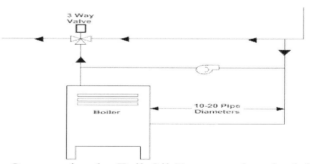

Once again, the Evil Oil Baron again raised fuel costs. This time citing the migratory patterns of pigeons are affected by global warming. Stingy Building Owner visited Stodgy Boiler Manufacturer who had the following suggestion.
"You need to install a blend pump and pipe it so that it will take hot supply water from the supply and mix it with the cooler return water to avoid thermal shock. The connection on the return

piping must be 10-20 pipe diameters away from the boiler to properly blend the water.

"The piping is four inch. We will have to completely re-pipe the boiler room" Stingy Building Owner supplied

"Well, it has to be 40 to 80 inches upstream to properly mix the water."

The Arrogant Wizard asked Stingy Building Owner for more money to install the new system.

"I have spent thousands of dollars already" Stingy Building Owner sighed.

"But, have you reduced your fuel costs?" The Arrogant Wizard asked

"Yes" he replied

After the latest installation was complete, the Arrogant Wizard handed a paper to Stingy Building Owner and said, "Oh, with all this new equipment, you will need an expensive service agreement."

Stingy Building Owner asked, "Will I save money with a service agreement?"

The Arrogant Wizard piled up some more rocks and smiled.

Remember, there are no shortcuts. You will not get 90% efficiency from a boiler designed to operate at 80% efficiency without sacrificing something.

On a small commercial boiler, a great way to reduce heating costs is to install a 2-stage thermostat. The first stage will energize the pump. If the building still requires heat after the pump is running, the burner will ignite. It is a good way to pull heat from the boiler without firing the burner. Another method is to install a time delay relay on the burner.

Primary Secondary
Many designers and manufacturers are using the Primary Secondary piping as their preferred method of piping boilers. There are several advantages to this type of piping arrangement.

Firstly, each boiler gets the proper flow since each boiler has a dedicated pump. Secondly, primary secondary piping has proven to be an

energy saver. The idle boilers are isolated so that the jacket and off time losses are limited. This meets the International Energy Code that calls for isolating the idle boilers. Lastly, it takes up less piping. Instead of return and supply piping in the boiler room, there is one loop.

61

Tales From the Field

We received a call from a Priest telling us that the boiler in the rectory would squeak every time it started. It was very loud and he was upset. Well, fearing eternal damnation, we went there right away. I accompanied the service technician because I could not figure out what was causing the squeaking and I was curious. The building had an old cast iron sectional steam boiler with an atmospheric burner. There were no moving parts.

When we arrived on the job site, the Priest was right. Each time the boiler started, there was a loud squealing that sounded like a loose belt. After some chin scratching, my service technician hit the flue with his hand when the boiler was off. The squealing started again. It turned out to be a nest full of bats living in the chimney. When the boiler started, it must have awakened the bats.

Did You Know...
The products of combustion produced when 1 cubic foot of gas is completely burned are 8 cubic feet of nitrogen, 1 cubic foot of carbon dioxide and 2 cubic feet of water vapor.

Chapter 7
Steam Boiler Sizing

Designing a replacement steam system is different from a hydronic system. To properly size a replacement steam boiler, you will need to add up all the radiation and piping. That is called the "Connected Load". In many instances, it is much larger than one that would be sized using a heat loss. Do not be tempted to install one according to the heat loss method. There will be many headaches and service calls. The boilers may never build up pressure if they are undersized resulting in high fuel costs and comfort complaints. Another common mistake is to size the new boilers according to the size of the old boilers. Many installers will not want to take the time to perform a study of the connected load. One thought is, What if the old boilers were the oversized or worse yet, undersized? You could be inheriting a huge headache. The following are some guidelines to help you size your new system. Add up all the radiators or heat emitters and the piping. Many of the old timers used to add an additional 33%-50% onto the size of the radiators for the piping load. This is to factor in the amount of heat required when the piping is cool. For instance, if the connected radiators equal about 400,000 Btuh, you may add anywhere from 150,000 to 200,000 Btu extra for the steam piping.

Can we convert it to hydronic?

With the increased acceptance of PEX tubing, the cost and labor it takes to convert the system from steam to hydronic is lower. In many buildings, the original engineer used a steam boiler to heat a steam to water convertor. See if the steam actually leaves the boiler room. Before considering a replacement steam boiler, consider converting it to hydronic.

Why Not Steam?

Steam has been classified as the "bad boy" of heating systems and I believe that it is an unjust title. I have seen steam systems that operate very efficiently. One of the most common points that the detractors bring up is how much heat it takes to change water to steam. Latent heat, or the amount of heat required to change one pound of water to one pound of steam is 970.4 btus. I like to remind them that the latent heat is simply borrowed from the boiler. The latent heat is given up into the space via a radiator or coil. Steam can travel at about 40 miles per hour without the use of a pump. A hydronic system pump utilizes a pump to distribute the hot water. In most instances, it operates 24 hours per day for the entire heating season. For example, a 2 horsepower pump will cost the building owner about $6.46 per day to operate. This would be based upon a Kwh cost of $0.1346. If you estimate a heating season of October 1 to March 31, the owner would pay $1,169.26 for the electrical consumption of the pump.

We installed a steam system for an old church and one of the members of the church was upset that the church chose to install a steam system rather than to convert the system to hydronic. We explained that converting the system to hydronic would have been more expense than the church could afford at this time. I told him that a well running steam system would lower their costs by about 20%. He scoffed and said, we will see. We installed the new boilers and replaced the trap inserts in the church. The church saw a 25% savings on their fuel costs.

Steam Pipe Insulation

Any steam boiler replacement should include insulation of the steam piping. This is a large heat loss and could affect the boiler sizing.

Steam Trap Retrofit

If you are replacing a steam heating system, this is a perfect opportunity to suggest the replacement of the steam traps. When replacing the steam traps, this will lower the fuel costs and increase the comfort of the system. Several companies sell steam trap inserts that make the replacement easy and cost effective.

According to SchoolFacilities.com, the following are the estimated life expectancy of steam traps:
Thermostatic Element 3 Seasons
Float Mechanism 6 Seasons

Controlling the Steam Boiler

Controlling the steam system is sometimes a tricky job. Several control panels that are on the market are designed for steam systems. They will typically have one sensor in the outside air and another that is located on the condensate pipe. The panel will cycle the boilers off once heat is established, as verified by the sensor on the condensate pipe. A drawback to this technology is that if steam traps are leaking through, the sensor could be fooled into thinking that the heat has been established and will shut off the boilers. If you are using one of these panels, I would suggest one that has a room sensor that will actually verify that steam is in the building. I have found that a simple programmable thermostat works almost as well as one of these expensive control panels in most instances.

Controlling a modular steam system may be a bit trickier. We have found that once the pipes are cold, you are better off starting all the boilers until the pressure is raised and then dropping them off. When you start only one boiler, there is a tendency of having carryover as the steam rushes out to the cool system and the single boiler cannot keep up with the load.

Radiator Sizing

Below are some charts that will help you to estimate the connected load. Dan Holohan from www.heatinghelp.com has some excellent resources for old steam systems.

Radiator Size Chart

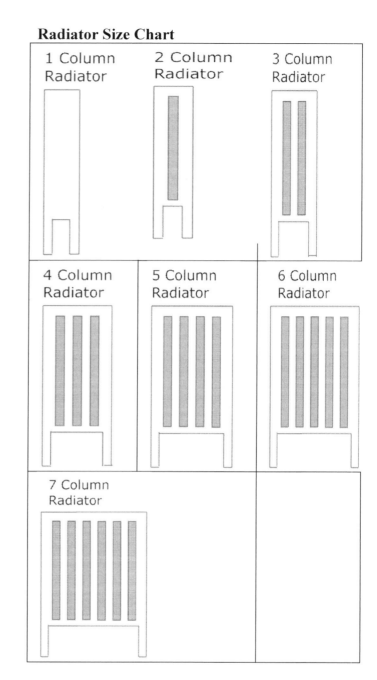

Cast Iron Radiator Ratings
Column Radiators

Column Radiators One Column

Height	Sq Ft / Section	Btuh
20"	1.5	360
23"	1.66	398
26'	2	480
32'	2.5	600
38"	3	720

Column Radiators Two Column

Height	Sq Ft / Section	Btuh
20"	2	480
23"	2.33	559
26'	2.66	638
32'	3.33	799
38"	4	960
45"	5	1,200

Column Radiators Three Column

Height	Sq Ft / Section	Btuh
18"	2.25	540
22"	3	720
26'	3.75	900
32'	4.5	1,080
38"	5	1,200
45"	6	1,440

Column Radiators Four Column

Height	Sq Ft / Section	Btuh
18"	3	720
22"	4	960
26'	5	1,200
32'	6.5	1,560
38"	8	1,920
45"	10	2,400

Cast Iron Radiator Ratings
Thin Tube Radiators

Three Tube

Height "	Sq. Ft. per Section	Btuh
20"	1.75	420
23"	2	480
26"	2.11	506.4
30"	3	720
36"	3.5	840

Four Tube

Height "	Sq. Ft. per Section	Btuh
20"	2.25	540
23"	2.5	600
26"	2.75	660
32"	3.5	840
37"	4.125	990

Five Tube

Height "	Sq. Ft. per Section	Btuh
20"	2.66	638.4
23"	3	720
26"	3.5	840
32"	4.33	1039.2
37"	5	1200

Six Tube

Height "	Sq. Ft. per Section	Btuh
20"	3	720
23"	3.5	840
26"	4	960
32"	5	1200
37"	6	1440

Seven Tube

Height "	Sq. Ft. per Section	Btuh
13"	2.625	630
16 1/2"	3.5	840
20"	4.25	1020

Communication Disorder

I had always considered myself a fairly good communicator until one fateful day in Philadelphia, PA. I was travelling with my brother and we pulled into a gas station. We were both tired from the trip. In this particular gas station, they still had an attendant pump your gasoline. This rather large man walked up to the car and said, "Help ya?"

I told him that I wanted to fill the tank. The attendant looked at me and asked, "Do ya'll eat chicken?" I thought it was an odd question but said "No Thanks. We just ate"

The attendant became upset and asked again, "Do ya'll eat chicken?" My brother smacked my shoulder and said, "Quit upsetting this guy." I again replied "No thanks. We just ate." With that, the attendant walked to my window and said "No Man, No! Do your oil need checkin?"

A young, newlywed couple were celebrating their first snowfall together. They decided to go for a walk in the newly fallen snow. The husband watched his wife putting her boots on and noticed that she pulled a bag from a loaf of bread onto each foot before placing her feet into the boot.

"Why did you do that?" he asked

"Do what?" she asked

"Why do you put the bread bag on your foot before putting them in your boot?" he asked

"I don't know. I have always done that. My mother would always save the bags from the bread for our boots as a child." she replied.

The next Sunday, the couple visited her mother's house for dinner. After dinner, the husband asked his new mother in law, *"Why did you make the kids put bread bags on their feet before putting on their boots?"*

The mother in law had a puzzled look on her face and replied, *"I don't know. My mother always did that."*

The husband shook his head and said, *"I don't understand. Was it to keep them dry? Or warm?"* The mother in law shrugged her shoulders.

A couple weeks passed and the grandmother came to the newlywed's house for dinner. After dinner, the husband poised the same question to the grandmother.

"Why did you make the kids wear bread bags on their feet under the boots?" he asked.

The grandmother smiled and patted the husband on the hand and said, *"As you know, I grew up in the Great Depression. Our old boots had holes in them and that was the only way to keep our feet dry when we played in the snow."*

Are we doing the same thing repeatedly without asking questions?

Chapter 8
Boiler Feed Sizing

In all steam boilers, the steaming rate is the same, regardless of style or construction. All steam boilers will convert the boiler water to steam at approximately one half gpm per 240,000 gross output Btu/Hr (D.O.E Heating capacity). A rule of thumb in the industry states that we need One gpm per 240,000 makeup to the boiler. This is so that the boiler does not run out of water.

In most steam boiler replacement projects, you will also need to replace the condensate tank with a boiler feed unit. When designing a boiler feed unit, the tank sizing is critical. If the tank is too small, this could waste treated water. For example, as the boiler steams, it relies upon the boiler feed unit to provide enough water so that the boiler will not shut off on low water. If the condensate is delayed returning from the building into the boiler feed tank, the tanks internal float will drop, introducing fresh, untreated water into the tank. When the condensate finally does return, the tank will flood, spilling water onto the floor. This can really decrease the life of the steam system because of the introduction of fresh water.

Sizing a boiler feed tank is not an exact science. Some formulas may be useful later in the chapter. To truly see how long it takes for the condensate to return to the boiler feed tank is to time it. This involves starting the system when the pipes are cool and timing how long it takes from when the boiler starts steaming to when the condensate returns to the system. In most instances, designers size the boiler feed storage tank for 10 to 30 minutes. This is the projected time it takes for the condensate to return from the system. Some manufacturers suggest a 50% safety factor when sizing a tank. In reality, there is usually only a small amount of cost difference between one tank size and the next. You are better off with a slightly larger tank than a slightly smaller one.

Condensate Tank

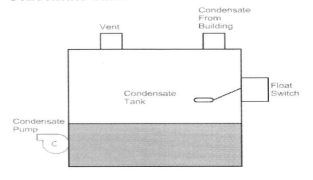

The condensate from the building will accumulate inside the tank. The condensate tank is smaller than a boiler feed tank, usually around 15-25 gallons. When the condensate water level raises high enough, the internal float switch energizes the pump and feeds water into the boiler, regardless of whether the boiler needs water or not. The pump will operate until the water level inside the tank drops low enough to open the float switch contacts. Any makeup water required by the steam system is fed through a level control located on the boiler. The makeup water for the system is fed directly into the boiler via a float-operated control. This could be problematic as cold water entering a hot boiler can cause many problems.

Boiler Feed Unit

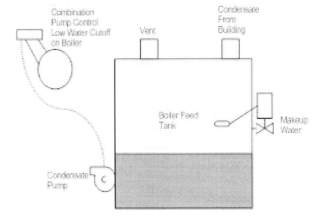

Condensate from the building will accumulate inside the boiler feed tank. The boiler feed tank is much larger than a condensate tank. When the condensate water level drops below the makeup water float, the valve attached to the makeup water float will open, feeding fresh water into the tank. As the tank water level rises, the makeup water float valve will close and stop feeding water into the tank. Operation of the pump is controlled by a combination pump control / low water cutoff, such as a McDonnell Miller #150, that is located on the boiler. As the boiler water level drops, the pump will be energized. The pump will operate until the boiler water level control is satisfied. Makeup water for the system is introduced into the boiler feed tank where it will be pre-heated.

Things to know
- BHP = Boiler Horse Power
- One Boiler HP = 34.5 lbs of steam/ hr from and at 212 degrees F.
- One gallon of water weighs approximately 8.337 pounds.
- 10 – 30 minutes storage time = rule of thumb for water storage for a boiler feed tank. This is the length of time it takes for condensate to return from the building.
- Multiply tank size by 1.5 for a safety factor.
- GPM = EDR x 0.000496
- GPM = BTU ÷ 480,000
- Pounds of Condensate/Hr = EDR ÷ 4
- To calculate boiler evaporation rate in gallons per minute GPM = BHP x .069 Example 100 BHP x 0.069 = 6.9 GPM

Calculate Storage Tank Sizing
Boiler Horsepower (BHP) x 34.5 / 8.337 lbs /60 minutes x Storage Time x (1.5) safety factor. This will give you a 50% larger tank for a safety factor.

For example 10 minute storage:
100 BHP x 34.5 / 8.337/ 60 x 10 = 68.9 gallons Multiply 68.9 x 1.5 (Safety Factor) = 103.35 Gallon Tank. If you choose a tank with a capacity greater than 103 gallons, that would be large enough for this project. This is based upon 10 minutes storage.

For example 20 minute storage:
100 BHP x 34.5 / 8.337/ 60 x 20 = 137.8 gallons Multiply 137.8 x 1.5 (Safety Factor) = 206.7 Gallon Tank. If you choose a tank with a capacity

greater than 206 gallons, that would be large enough for this project. This is based upon 20 minutes storage. A larger storage tank may be required if the condensate is slow to return. A safety factor is required because of possible system deficiencies and the fact that some of the tank is below the pump inlet.

A couple other rules of thumb for tank sizing are:
- Boiler evaporation rate in GPM x 20 = Tank Size. (This is based on 10 minute storage)
- Boiler evaporation rate in GPM x 40 = Tank Size. (This is based on 20 minute storage)
- Boiler evaporation rate in GPM x 60 = Tank Size. (This is based on 30 minute storage)
- One gallon of storage for each boiler HP for a small building. Two gallons of storage for each boiler hp for a larger building.

Boiler Feed Water Pump Sizing
To calculate the pump capacity, you will need to know the evaporation rate of the boiler. You should then add a 50% to 100% safety margin.

To calculate the boiler evaporation rate, please use the following formula:
Evaporation rate =
Boiler HP x 34.5 / 8.337(lbs) / 60 (minutes)

For Example:
BHP x 34.5 / 8.337 / 60 minutes
100 BHP x 34.5 / 8.337 / 60 = 6.89 Gpm (Evaporation Rate) x 1.5 (50% Safety Factor) = 10.33 Gpm Pump

The following are some rules of thumb for boiler feed pump sizing
- 1/10 gpm per Boiler HP or BHP E.g. 100 hp = 10 gpm
- 2 times boiler maximum evaporation rate or 0.14 GPM per boiler HP for intermittent operation.

- 1.5 times boiler maximum evaporation rate or 0.104 GPM per boiler HP for continuous operation.

Pump Discharge Pressure

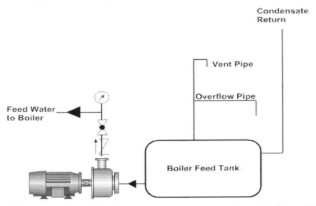

The pump discharge pressure should be 3% higher than the relief valve setting plus pressure drop. Always install a valve on the discharge of the pump to limit the feed water pressure.

For example:
15 psig relief valve setting and 2 # pressure drop 15 x 1.03 + 2 = 17.45 #. Your pump will have to have a discharge pressure of at least 17.45 pounds.

Therefore, our system will consist of the following

Tank Size -	103 gallon (10 Minutes)
	Or
	206 gallon (20 Minutes)
Pump GPM -	10 GPM
Pump Discharge	17.45 #

Another Sizing Option
100 BHP Gross Output = 3,347,200 Btuh
3,347,200 / 970 Btu = 3,451 Lbs of steam evaporated per hour.

The evaporation rate is 3,451 / 8.33 = 414 gallons per hour (gph) or 6.90 gallons per minute GPM. Multiply 6.90 x 1.5 (50%) safety factor) for the pump GPM size.

Tank size 414 / 3(20 minutes before condensate begins to return) = 138 gallons. Multiply x 1.5 for safety factor. This equals tank size of 207 gallons.

Condensate Tank & Pump Sizing
Evaporation rate x 3 = Pump GPM required
Pump GPM x 1 = Tank sizing
Please note that most new boilers require a boiler feed instead of a condensate tank.
If you want a chart that calculates the sizing, please see below.

	Boiler Feed Tank Sizing			
Boiler HP	10 Minutes	20 Minutes	30 Minutes	Pump Sizing GPM
20	21	41	62	2.07
30	31	62	93	3.10
40	41	83	124	4.14
50	52	103	155	5.17
60	62	124	186	6.21
70	72	145	217	7.24
80	83	166	248	8.28
90	93	186	279	9.31
100	103	207	310	10.35
110	114	228	341	11.38
120	124	248	372	12.41
130	134	269	403	13.45
140	145	290	435	14.48
150	155	310	466	15.52
160	166	331	497	16.55
170	176	352	528	17.59
180	186	372	559	18.62
190	197	393	590	19.66
200	207	414	621	20.69
250	259	517	776	25.86
300	310	621	931	31.04
350	362	724	1,086	36.21
400	414	828	1,241	41.38
Tank & pump sizing based upon 50% safety factor				

Pump vs. Solenoid Valve

Some installers prefer to use one boiler feed pump per boiler. An advantage to this is that each boiler will have its own pump. One drawback to this option is that if one pump fails, you will not be able to run that boiler. The other disadvantage is that you will need to run one pipe from the boiler feed pump to each boiler. If there were three boilers, you would need three pipes and three pumps.

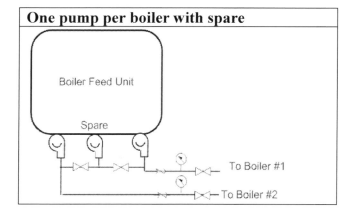

One pump per boiler with spare

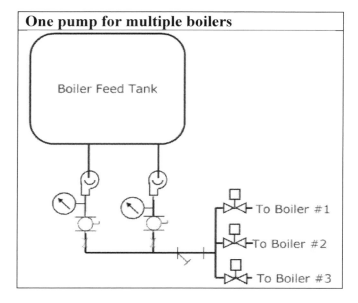

One pump for multiple boilers

Some designers would prefer to use feed-water valves instead. The advantage to this is that the installer only has to install one pipe to feed all the boilers. The disadvantage is that there could be a lag in the response time. If you have a small boiler, it could "trip" the low water cutoff by the time the control calls for water and the pump energizes.

Strainers should be installed on all piping before the solenoid valve to keep debris from entering the valve. In some instances, you may have to operate the pump continuously to make sure that the water gets to the boiler before the low water cutoff shuts off the burner.

Backup Pump

Some designers prefer a backup pump in the event of a failure. This can be accomplished by installing a duplex boiler feed unit and using the feed water valves.

Steam Trap Primer
The following was provided by Engineeringtoolbox.com

Inverted Bucket Steam Trap

The inverted bucket is the most reliable steam trap operating principle known. The heart of its simple design is a unique leverage system that multiplies the force provided by the bucket to open the valve against pressure. Since the bucket is open at the bottom, it resists damage from water hammers, and wearing points are heavily reinforced for long life.

- intermittent operation - condensate drainage is continuous, discharge is intermittent
- small dribble at no load, intermittent at light and normal load, continuous at full load
- excellent energy conservation
- excellent resistance to wear
- excellent corrosion resistance
- excellent resistance to hydraulic shocks
- vents air and CO_2 at steam temperature
- poor ability to vent air at very low pressure
- fair ability to handle start up air loads
- excellent operation against back pressure
- good resistance to damage from freezing
- excellent ability to purge system
- excellent performance on very light loads
- immediate responsiveness to slugs of condensate
- excellent ability to handle dirt
- large comparative physical size
- fair ability to handle flash steam
- open at mechanical failure

Thermostatic Steam Traps

There are two basic designs for the thermostatic steam trap, a bimetallic and a balanced pressure design. Both designs use the difference in temperature between live steam and condensate or air to control the release of condensate and air from the steam line.

In an thermostatic bimetallic trap it is common that an oil filled element expands when heated to close a valve against a seat. It may be possible to adjust the discharge temperature of the trap - often between $60^{\circ}C$ and $100^{\circ}C$.

This makes the thermostatic trap suited to get rid of large quantities of air and cold condensate at the start-up condition. On the other hand the thermostatic trap will have problems to adapt to the variations common in modulating heat exchangers.

- intermittent operation
- fair energy conservation
- fair resistance to wear
- good corrosion resistance
- poor resistance to hydraulic shocks (good for bimetal traps)
- do not vent air and CO_2 at steam temperature
- good ability to vent air at very low pressure
- excellent ability to handle start up air loads
- excellent operation against back pressure
- good resistance to damage from freezing
- good ability to purge system
- excellent performance on very light loads
- delayed responsiveness to slugs of condensate
- fair ability to handle dirt
- small comparative physical size
- poor ability to handle flash steam
- open or closed at mechanical failure depending of the construction

Float Steam Traps

In the float steam trap a valve is connected to a float in such a way that a valve opens when the float rises.

The float steam trap adapts very well to varying conditions as is the best choice for modulating heat exchangers, but the float steam trap is relatively expensive and not very robust against water hammers.

- continuous operation but may cycle at high pressures
- no action at no load, continuous at full load
- good energy conservation
- good resistance to wear
- good corrosion resistance
- poor resistance to hydraulic shocks
- do not vent air and CO_2 at steam temperature
- excellent ability to vent air at very low pressure
- excellent ability to handle start up air loads
- excellent operation against back pressure
- poor resistance to damage from freezing
- fair ability to purge system
- excellent performance on very light loads
- immediate responsiveness to slugs of condensate
- poor ability to handle dirt
- large comparative physical size
- poor ability to handle flash steam
- closed at mechanical failure

Thermodynamic Disc Steam Traps

The thermodynamic trap is an robust steam trap with simple operation. The trap operates by means of the dynamic effect of flash steam as it passes through the trap.

- intermittent operation
- poor energy conservation
- poor resistance to wear
- excellent corrosion resistance
- excellent resistance to hydraulic shocks
- do not vent air and CO_2 at steam temperature
- not recommended at low pressure operations
- poor ability to handle start up air loads
- poor operation against back pressure
- good resistance to damage from freezing
- excellent ability to purge system
- poor performance on very light loads
- delayed responsiveness to slugs of condensate
- poor ability to handle dirt
- small comparative physical size
- poor ability to handle flash steam
- open at mechanical failure

Chapter 9
How Many Boilers?

When replacing a heating system consisting of a single boiler, now is the time to educate the client on different strategies. The client may think that one boiler worked well for years. We will compare three different strategies for heating a building with a design load of 800,000 Btuh.

Use One Boiler

The advantage to installing only one boiler would be the installation cost. There are a couple disadvantages to this option.

Firstly, there would be no backup in the event of a parts failure. Your client would be without heat until the part is obtained and installed. *Have you noticed that no one seems to keep parts in stock now?*

The second is increased fuel costs. The original single boiler was probably installed when fuel was inexpensive. The boiler would have been sized for the design temperature of the locale plus a safety factor. This means it will be grossly oversized for the majority of the winter. For example, our building design load is 800,000. Assuming an 80% efficient boiler, we would need a boiler rated at 1,000,000 Btuh. In most buildings, our boiler will have a safety factor of anywhere from 10-40% .That means that our boiler will be sized at 1,100,000 to 1,500,000 Btuh. In this instance, we will say that the boiler is sized conservatively and is 1,100,000 Btuh. The jacket loss from the boiler would be anywhere from 2-5% of the rating. That equals 22,000 to 55,000 Btuh will be lost from the boiler each hour into the boiler room.

The last disadvantage is that the boiler will cycle constantly in the warmer weather, increasing the likelihood of a break down. According to the International Energy Conservation Code, a single boiler with a rating above 500,000 Btuh will require a multistage or modulating burner.

Install Two Boilers

For years, this was the preferred method by most designers. Heating systems were designed for comfort and not necessarily efficiency. The designers worst fear was that the owner would call and say that the new boiler they designed was down and the building had no heat. Let us assume that our designer, who was clad in a leisure suit, pocket protector and platform shoes

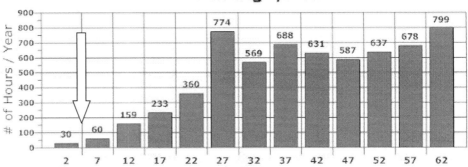

Hourly Temperatures Pittsburgh, PA

# of Hours / Year vs. Outdoor Temperatures

in the 1970's, used the traditional design of the day that included two large boilers. The heat load of the building was split between the two boilers. Since heating costs were low, each boiler would be sized for between 66 %(2/3) to 75 %(3/4) of the estimated load. In the event of a boiler malfunction, one boiler could provide heat to the building. In some instances, such as hospitals, they require 100% redundancy. That means that each boiler would be sized to handle 100% of the load in the event of a breakdown.

In this example, we will use 3/4 or 75% load factor for each boiler. This was the most common sizing percentage. As stated above, we use 800,000 Btuh as our building heat loss at the design temperature. For two boilers each sized at 75% of the load, we will require two boilers each rated at a net output of 600,000 Btuh. At 80% efficiency, each boiler input will be 750,000 Btuh. If the boilers were installed in the traditional way, the water would flow through the boilers all the time. The jacket loss from the boilers into the boiler room would range from 2-5% of the ratings for each boiler. That would equal 30,000 to 75,000 Btuh losses each hour into the boiler room.

At that time, this was the prudent thing to do but consider this. With two boilers sized at 75% of the load, the heating plant would be oversized by 50% on the coldest days. As the outdoor temperature rises, the oversized percentage would be even greater. Please see chart below that details the oversized percentage. Another factor to consider is the heating hours per season. The design temperature occurs 2 1/2% of the winter. Based upon 4,000 heating hours per year in my locale, the building would be at the design temperature for 100 hours per year.

Boiler Plant Over sizing vs. Outside Air 2 Boilers each sized at 75% of load		
OA Temp	% of Building Heat Loss	Heating Plant Oversized %
5 Design Temperature	100%	150%
10	90%	167%
15	80%	188%
20	70%	214%
25	60%	250%
30	50%	300%
35	40%	375%
40	30%	500%
45	20%	750%
50	10%	1,500%

Install Modular Boilers

Now, consider a boiler room with four modular boilers, each sized at 25% of the load, piped with a primary secondary main. Each boiler would be rated at 250,000 Btuh. In the event of a boiler malfunction, the building would still have 75% backup just like the above example of the two boilers. However, your heating system would be

at least 10% smaller than a system with one boiler and 33% smaller than the system with two boiler sized at 75% of the load. In addition to having a smaller carbon footprint, you would have a smaller flue, chimney and combustion air louver. On a day when only one boiler was firing, your jacket loss would be 5,000 Btuh or 2% of one boiler. The other boilers would be isolated because they were piped in a primary secondary arrangement. In addition, the jacket loss would only occur when the boiler was firing. Compare that loss of 5,000 Btuh with a modular boiler system to the other jacket losses and you would see significant savings. The jacket loss would be 88-91% lower than a single boiler and 84-94% lower than two boilers. Modular boilers, piped in a primary secondary arrangement, will save you money.

Modular boilers are becoming the staple of the industry now because of several advantages. The seasonal efficiency is much higher than with a single boiler or two large boilers. The maintenance is easier because the boilers are smaller and less sophisticated. There is back up in the event of boiler malfunction.

Modular Boilers Courtesy Triad Boiler Systems

Size Does Matter

According to an EPA study entitled, "Guidelines for Industrial Boiler Performance Improvement", a boiler is most efficient when operating with a 50% to 80% load range. In other words, operating a boiler at low fire is the least efficient way to operate the boiler. In addition, the study found that boiler efficiency drops dramatically if operated below the 50% load factor.

Another interesting part of the study was that operation of the burner with greater than a 2:1 turndown is not necessary for fuel-efficient operation.

The study also suggested that a fully modulating boiler operating continuously at 37% of the full load would be less efficient than a boiler operating on-off in the 50-80% firing range.

Tales From the Field

I was visiting a jobsite in rural West Virginia. As I was talking with the director of maintenance, I noticed yellow ribbons attached to the windows.

"It's a nice gesture to support the troops" I said as I pointed to the ribbons.

The guy looked at me as if I was crazy.

I pointed to the ribbons and said "I noticed the yellow ribbons on the window and thought that was a good thing to support the troops."

The guy gave me the same look and said,

"It's to keep the damn birds from flying into the windows"

Did You Know…
The American Society of Mechanical Engineers ASME was founded in 1880

Chapter 10
Modulating or On Off Burners?

To reduce the cycling of the oversized boiler, some designers specify multistage or modulating burners. According to the International Energy Conservation Code, a single boiler with a rating above 500,000 Btuh will require a multistage or modulating burner. There are three choices for multi-stage burners, Low High Off, Low-High-Low, and Modulating.

Low-High-Off Burners
On a call for heat, the pilot light is lit and verified. After the pilot is established, the Low High Off burner will start at low fire. It then will move to high fire until the end of the heat cycle when it will shut off.

Low-High-Low Burners
On a call for heat, the pilot light is lit and verified. After the pilot is established, the Low High Low burner will start at low fire. It then will move to high fire. As the burner gets close to the temperature or pressure set point, the burner will cycle to low fire. In an attempt to maintain the set point, the burner will fire at either low or high fire until the call for heat is complete. According to Honeywell in their book entitled, Flame Safeguard Controls, a Low High Low burner can use 10-15% less fuel than a modulating burner.

Modulating Burner

Courtesy of Iron Fireman

On a call for heat, after the pilot is established, modulating burners start at low fire. It will then move to any point between low and high fire to main-tain the set point. The modulating set point is usually just below the operating control set point. For example, the boiler operating control will be set for 180^0F. The modulating control may be set for 175^0F. When the temperature reaches 175^0F, the burner will drop to low fire. Most of the smaller commercial modulating burners will have a turn down of three to one. A three to one turndown means that the burner will drop to 33% of the high fire rate at low fire. Some larger burners will have a ten to one turndown. That means that the burner will drop to 10% of its firing rate. Be careful of a high turndown as it could cause the flue gases to condense inside the boiler and stack.

"Modulating burners save money" For years, we have heard that and, in a way, it is true. If you consider our previous example, you would have two boilers, each sized for 75% of the load. The system circulating pumps would continuously flow through the boilers, even when the burners were not firing. In milder weather, the on-off burners would cycle continuously. Each time the burners cycled, you would have pre and sometimes post purge losses. On a call for heat, the burner blower starts and will operate anywhere from 30 seconds to several minutes. This is to "purge" the boiler combustion chamber of any combustible gases. The length of pre-purge will be as long as it takes to have four complete air changes inside the boiler prior to any fuel being introduced into the boiler. The older boilers used to have seven air changes. The blower takes the air from the boiler room and sends it up the stack, taking heat with it as it goes. This is part of what is called "cycling costs". On larger boilers, there is sometimes a post-purge cycle where the fan will operate after the fuel is shut off.

One of the drawbacks to modulating burners is that it is almost impossible to adjust the linkages for peak efficiency throughout the entire firing range due to the inherent sloppiness of the linkages. The linkages connect the combustion air to the fuel supply to the burner for efficient and safe firing. The time it takes to adjust the fuel to air linkages is much longer than an On Off or Low High Low burner. Honeywell recognized this flaw and developed linkage-less controls. Honeywell contends that burner linkages result in a decrease of 4-6% in boiler efficiency.

Another drawback to modulating burners is that the run time of the boiler and burner are extended. When the run time of the burner is extended, the heat loss from the boiler into the room is also extended. This is called "jacket loss" and it ranges from 2-5% of the boiler's rating. For example, if you have two boilers each rated at 1,000,000 Btuh, you would lose between 40,000 and 100,000 Btuh per hour from the boiler into the room at any time the water inside the boiler is warm. (That is why the IECC code calls for isolation of the unfired boilers.). Let us look at a condition where only one boiler was firing at low fire. The low fire position is 33% of the firing rate. In this instance, our input would be 330,000 Btuh at low fire. The boilers are still losing the same amount of heat as when it was at high fire but now your jacket loss is between 12-30% of the firing rate. Your 80% efficient boiler just dropped to between 50-68% efficient. This does not factor in the sloppiness of the linkages, which could lower your efficiency by another 4-6%.

The burners are designed to have turbulence to mix the fuel and air properly. At low fire, some burners do not have the proper turbulence and the fuel and air do not mix properly. This could lead to carbon monoxide production inside the boiler.

Should Your Client Buy The More Expensive Boiler?

I have a theory that most clients will always choose the least expensive option if they are uninformed of what they are purchasing. Let us look at a scenario that could happen when looking at a boiler replacement project. Your proposal should include two or three options. I like the Good, Better, Best approach for my proposals. In this scenario, we will look at two different options. The first is a standard boiler and the second is a higher priced boiler that will cost $5,000.00 more than the standard boiler. The estimated savings are about $2,500.00 per year.

At first glance, it appears to be a decent value. They would have a return on investment in two years. In this tight economy, when every penny counts, the client may balk at spending the extra $5,000.00. That is where you need to understand the dynamics of commercial real estate financing.

If the owner chooses the less expensive option and puts that $5,000.00 in a savings account for twenty years, earning 2% per year compounded monthly, he or she would earn $2,456.64 in interest, bringing the $5,000.00 investment to $7,456.64 after twenty years.

Now, if the owner puts that same amount into a more efficient boiler that costs $5,000.00 more and will save $2,500.00 or about $208.00 per month, what is the impact of that? The future value of the deposits i.e. energy savings would be $208.00 per month for 20 years @ 2% interest. This would equal $61,317.74 after twenty years. In addition, there is a rule of thumb in the real estate industry that states that the building value is increased by $10.00 for every for every dollar in profit. In our above example, we estimate that the building heating costs will be reduced by $2,500.00 per year. This increases the building value by $25,000.00. Therefore, our initial investment of $5,000.00 will throw $86,317.74 back to the owners for the 20-year life expectancy of the boilers. I would rather have $86,000.00 than $7,456.64. This does not factor in the increased fuel costs.

Does Night Set Back Really Save Money?

According to a York Shipley article entitled, "Minimum Firetube Boiler Flue Gas Stack Temperatures", it states the following, "Day/Night set-back operating systems amplify the potential boiler damage as low operating temperatures, and will use 25% to 50% additional fuel per day than a conventional primary/secondary loop system operating around the clock." The article also suggests that low heating system temperature operation will result in premature pressure vessel failures, with repair costs exceeding fuel savings by a minimum factor of ten. The article also recommends the following minimum flue gas temperatures

Natural Gas 265^0F plus $1/2^0$F for each foot of stack or breeching, including horizontal and vertical runs

#2 Fuel Oil 240^0F plus $1/2^0$F for each foot of stack or breeching, including horizontal and vertical runs.

Did You Know...
There are roughly 163,000 boilers in use in the US.

Did You Know...
A cubic inch of water evaporated under ordinary atmospheric pressure (14.7 psig) will be converted into approximately one cubic foot of steam.

Did You Know…
A burner should operate for 15 minutes before adjusting the fuel to air ratio to allow the flame to stabalize.

Did You Know…
That all steam boilers will lose water to steam at a rate of ½ gpm for each 240,000 Btuh.

Did You Know…
The first commercial steam engine was patented by Thomas Savery in 1698.

Chapter 11
Condensing Boilers?

Condensing boilers are like that high maintenance car we all had when we were young. It looks good but requires much care. Should we install condensing boilers? This is a decision that will be based upon a couple factors. The first is the installation budget. Condensing boilers tend to be more expensive to install than traditional boilers. Many require stainless steel venting. Most require a separate vent from the outside for combustion air.

A second factor is maintenance. The condensing boiler will require more maintenance than a traditional boiler.

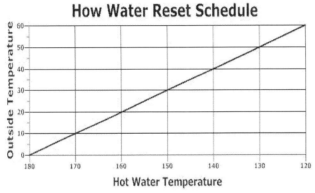

The last and most important factor is the medium that you are heating. If you are heating a medium that requires an inherently low temperature like snowmelt, in-floor radiant heat or water source heat pumps, a condensing boiler is ideal. If you are connecting to a traditional heating system, you may want to rethink a condensing boiler. What most people do not realize is that a condensing boiler will only condense when the water temperature is below 140 degrees. When the water temperature is above 140 degrees F, the condensing boiler will operate like a traditional boiler. That means that the more expensive, high efficiency, condensing boiler will have an efficiency rating in the mid to upper 80% range when the water is warm. It is kind of like getting Clark Kent when you wanted Superman. The worst part is that the condensing boilers will not

hit 90% efficiency until the water temperature is at 100^0F or below I like seeing the brochures for these Über efficient boilers with claims that they are 99% efficient. What they say in small font is that you will need to have water cooler than 60^0F entering the boiler to achieve 99% efficiency. In my area, a pipe carrying 60^0F or 70^0 water will not provide much heat to a room that is at 70^0F trying to reach 72^0F.

Most older heating systems were designed to operate at 180^0F at the design temperature. If you estimate a linear reset schedule, the average boiler water temperature will not drop below 140 degrees until the outside temperature is above 40 degrees F. In this example, our linear reset would go from 180 degrees F at 0 degrees outside temperature to 120 degrees at 60 degrees F outside temperature. If you reference the hourly temperatures chart below, you will see that the "condensing" boiler will not condense for a large portion most of the winter. In Europe, the designers are over sizing the heat emitters. The heat emitter may be something like radiators or fin tube radiation. By over sizing the heat

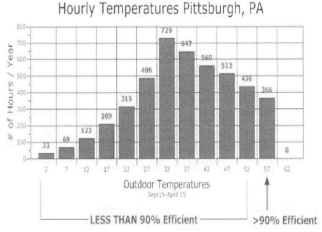

emitters, they can operate the boilers at a much lower temperature and still have adequate heat for the building.

Is that feasible to do here? Consider a hypothetical building with a heat loss of 500,000 Btuh. In this building, we have Slant Fin model 90-21 fin tube radiation that is rated for 1,007 Btuh at 180^0 F water. At that temperature, our building would require 496 feet of radiation to heat the building at the design temperature. We want our new boiler to be fully condensing in the winter so we decide to size the radiation for 120^0F. The Btuh rating for the same Slant Fin tubing at 120^0F is 380 Btu's per linear foot. To heat our building using 120-degree water, we would need 1,315 feet of radiation. That is more than **triple** the amount of radiation that is in the building. It may be physically impossible to add that much more radiation.

The last consideration as to whether to install traditional boilers or standard boilers is the estimated life expectancy. The standard boilers will have a much longer life expectancy than a condensing boiler. In Europe, where condensing boilers have been used for a far longer time, they estimate a life expectancy of 10-15 years, according to UK Energy Saving. The Chartered Institution of Building Service Engineers (CIBSE) estimates a 15-year life of condensing boilers. This is about half the estimated life of a standard boiler.

Condensing boilers do have a place in our industry but they do not fit into every application. For example, a designer wanted condensing boilers for a college. The system was very complex as the designer was using the boilers to provide heat for space heating, domestic hot water and to heat a swimming pool. He would use a water to water heat exchanger to heat the pool and domestic hot water. What he did not realize was that the domestic hot water required 140 °F and to get that temperature, the boiler had to provide 160 °F water to the heat exchanger. Since the water temperature was that warm, the condensing boiler rarely condensed. The existing heating system in the building required 180 °F at the design temperature. A standard efficiency boiler would have had the same results. The designer had the owner spend about 20% more for condensing boilers.

Savings
In spite of the increased care that they require, condensing boilers can still reduce heating costs over traditional boilers.

Is a Hybrid System Right For the Project?
Even with these limitations to condensing boilers, they are being installed at a record pace for one reason; *lower heating costs.* Heating costs are gobbling up more and more of the operating budgets in commercial buildings.
What if we were able to combine the benefits of both types of systems; the reduced maintenance and longer life of a standard boiler and the energy savings of a condensing boiler? That is exactly what a hybrid heating system does. It combines one or more standard efficiency boilers with one or more condensing boilers. The standard efficiency boilers will be the lead boilers when the outside temperature is below 32°F. The condensing boilers will take over the lead when the outdoor temperature is above 32°F. In this way, you get longer life, lower maintenance costs, lower installed cost and increased system efficiency. In addition, the life expectancy of the condensing boiler is greatly extended because the run time is only half of the winter.

What is Boiler Efficiency?
Boiler efficiency is a vague term that compares how much heat is delivered to the building to how much fuel was consumed. There are several types of efficiencies when discussing boilers.

Combustion Efficiency
This is the efficiency of the burner during the burn cycle. It is a calculation of the heat input to the burner less the unburned fuel and excess air in the stack of the boiler. It is measured with a combustion analyzer. This only measures the efficiency of the boiler when the burner operates.

Boiler Efficiency
This is sometimes known as Fuel to Steam Efficiency. It is a more accurate method to calculate the efficiency of the boiler but really does not give you the true efficiency of the

system. It is the amount of heat delivered to the building or process compared to the actual fuel input. Boiler efficiency can vary due to various factors such as

- scale build up
- soot accumulation
- water flow through boiler
- dirty tubes

To accurately measure boiler efficiency, you would need to measure the Btu output vs. the Btu input.

Seasonal Efficiency

This is the most difficult to measure because it is affected by

- Cycling loss
- Jacket loss
- On and Off times of the boiler

This is what the boiler costs to operate over an entire heating season. This is the true way of measuring boiler efficiency.

Sizing a Condensing Boiler

Be careful when sizing a condensing boiler. Many of the condensing boilers will show an efficiency rating in the high 90% on their brochures. In reality, the efficiency drops off as the water get warmer. To exemplify this, consider a 1,000,000 Btuh boiler rated at 95% efficient. You assume that that the boiler will provide 950,000 Btuh. When you look at the fine print of the boiler brochure, it shows that the efficiency of the boiler drops to 85% when the water temperature is above 160^0F. You need to factor that in when sizing the new system. So, when your building requires the most heat, the condensing boiler provides the least heat

Bin Temperatures

The following is a graphical illustration of the Hourly Temperature Occurrences in two cities, Charleston, WV and Pittsburgh, PA. The temperature occurrences are a great way to illustrate the hours in a typical year. They are available on line from the NOAA, National Oceanic and Atmospheric Administration. They are a great sales tool for selling boilers.

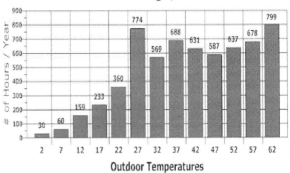

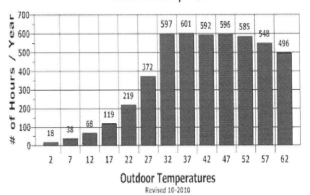

Did You Know…
One cubic foot of steam exerts a mechanical force equal to that needed to lift 1,955 pounds one foot.

Did You Know…
The average life of properly treated condensate piping is between 10-20 years.

Did You Know…
The low water cutoff is the leading mechanical cause of boiler accidents.

Did You Know…
Boilers with more than 500 sq. ft of heating surface require two or more relief or safety valves

Chapter 12
Other Boiler Room Considerations

Combination Boilers

Some boilers use an internal or external heat exchanger to heat domestic hot water as well as providing space heating for the building. This provides some advantages over traditional water heaters. The first is reduced scale.

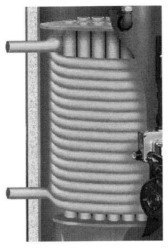

Combination Boiler
Courtesy Triad Boiler Systems

Since the boiler water heats the domestic water via a water-to-water heat exchanger, it is now an indirect water heater. That means that the scale build up, common for traditional water heaters, is virtually eliminated. Scale usually forms where the flame hits the metal because fresh water is always introduced to the system.

The second benefit to a combination boiler is that the combustion air requirements for the building are reduced. A further benefit is a smaller carbon footprint. Because the space heating boiler is heating the domestic water, you will not need another fuel-fired vessel to heat the water. Lastly, you would have one less flue to install.

Dual Fuel Burners

Some facilities require dual fuel burners as a backup fuel source in the event of a disruption in their natural gas supply. The most common dual fuel combination is natural gas and #2 fuel oil. If the client requires 100% backup, check to make sure that your burner is truly dual fuel. Many combination burners use a natural gas pilot to light the fuel oil. If that is the case, the client will not have heat in the event of a disruption in the natural gas. In some locations, you may need to connect a propane tank to the pilot in the event of a gas outage. Before doing this, check with the burner manufacturer to see what would be required to fire the pilot on propane gas. The Btu content of propane is higher than natural gas. They may require a separate pilot assembly or an orifice.

Front End Loading

When I first heard this term, I thought what does breast augmentation have to do with boilers? It turns out to be a unique idea where the budgets are tight. The installer keeps one of the older, large boilers in place. The installer will install from one to several smaller boilers to heat the building when the outside temperature is above 32 degrees F. Below that temperature, the large existing boiler is the lead boiler. In this way, the client gets the benefit of an efficient heating system without the high initial cost.

Replacement Burners

Another way to save money is to replace the burners with new more efficient burners. A 10% reduction in fuel costs is not uncommon with replacement burners without a large financial outlay.

A Money Saving Idea

A school district in our area had some of these old warm air furnaces. The heat exchangers were questionable and they were looking at spending a large amount of money to replace them. We worked with the school district and their contractor to offer a cost effective alternative. We provided boilers and the contractor installed a hydronic

coil inside the furnace. The customer had a renewed heating system and lowered their heating costs by 25%.

Did You Know…
ASHRAE was started in 1894 and currently has over 51,000 members

Did You Know…
The first electric thermostat was invented in 1883 by Warren S. Johnson, founder of Johnson Controls and a school teacher.

Did You Know…
The relief valve was patented in 1682 by a Frenchman named Denis Papin

Did You Know…
Most dissolved solids in a steam boiler are 4-6" below the water line?

Chapter 13
Choosing the Boiler Construction

A Deciding Factor...
In some instances, boiler selection is dictated by the available space in the boiler room. If you only have room for one boiler, you may be forced to install just one.

Boiler types
"What kind of boilers are the best?" is a question I hear frequently during my boiler classes. My response is "What manufacturer of car do you prefer?" I may prefer one brand or style but it does not mean that others are better or worse. They all do the same thing; heat water. I think there a couple of aspects that should factor into your decision.

In my years in the industry, I have learned a couple indisputable facts. The first is that the best boiler in the world will not operate correctly if it is installed wrong. Conversely, if the worst boiler in the world is installed and maintained correctly, it will last almost as long as the best. The second is that all boilers require maintenance. If they are not maintained, they will not last.

Factory Support - There is nothing worse than having a problem job and feeling as if you are alone in the battle. This is especially true if the owner is breathing down your neck or the threat of a lawsuit is looming. This includes the support from the manufacturer's representative.

Maintenance- Are the boilers easy to maintain? Sometimes, the more complex the system is, the less maintenance that is performed. If the equipment intimidates the maintenance person or service technician, they will not service it. If the system is not maintained, it will not last.

Parts Availability – Nothing angers an owner more than when they find out that the parts they need to get heat are not available. If the boiler you choose has proprietary parts, you should suggest some common replacement parts such as ignition electrodes or flame sensors.

There are several types of boilers available for replacement in your building. This may give you some pros and cons of each boiler type.

The following are non condensing boilers.

Copper Boiler	
Advantages	**Disadvantages**
Weighs less than a cast or steel boiler so it easier to handle.	Parts are proprietary and only available from the factory.
Most are factory tested to reduce jobsite problems.	Boiler uses a flow switch instead of a low water cut-off so flow through boiler is crucial.
Can handle lower temperature than cast iron or steel boilers.	The thermal mass of copper boiler is small so it tends to cycle quickly, reducing seasonal efficiency.
Has lower maintenance cost than a cast iron boiler according to ASHRAE	Higher maintenance costs than steel fire tube boiler according to ASHRAE

Cast Iron Boilers	
Advantages	Disadvantages
Is available in sections that will fit through most doors. This allows for field erection of the boilers.	The boiler is field erected by the "Low Bidder" and requires extra time on jobsite. Labor in the field is usually more expensive than labor in the boiler factory.
Replacement sections are available in the event that one is defective.	Highest maintenance costs of any boiler type according to ASHRAE.
The cast iron boiler has a higher thermal mass, which results in less cycling than a copper finned boiler.	Parts are proprietary and only available from the factory. Since the larger boilers are field erected, there is no assurance that the boiler is installed according to the factory.
The cast iron boiler is a rugged boiler that has a long life expectancy.	Some cast iron boilers had a problem with flue gases leaking from inside the boiler between the sections into the boiler room.
Cast iron boilers have been in the field for many years and are a proven item.	Boiler reset is limited in these type boilers, as they typically have to have entering water temperatures higher than 140 degrees F.
	Cast boilers are difficult to clean on both the fire side and water side.

Packaged Steel Boilers	
Advantages	Disadvantages
Factory assembled and tested so that job site problems are minimized.	More expensive than copper boilers.
Lowest maintenance costs according to ASHRAE.	May not fit through access door
Parts are available at most supply houses.	Is susceptible to thermal shock
Steel boilers have been in the field for many years and are a proven item.	Limited reset schedule
Some packaged boilers will fit through doors.	A certified welder must do any welds on the boiler.
Some packaged boilers are factory wired to reduce installation times	
Has a higher thermal mass, which results in less cycling than a copper finned boiler	

Condensing Boilers

Choosing whether to install a condensing boiler or a standard efficiency boiler is difficult. The following chart will help you look at the pros and cons of each type.

Condensing Boiler Compared to Non Condensing Boilers	
Advantages	**Disadvantages**
Easier to handle than a traditional boiler.	Lower life than a standard efficiency boiler.
Less likely to be subjected to fluctuating draft conditions in chimney.	Requires more maintenance than a traditional boiler.
Less likely to cause a negative condition in a boiler room.	Higher installed cost than a traditional boiler.
Can be installed in same room as an air conditioning unit without a refrigerant monitoring system as long as combustion air is vented from outside.	Most require combustion air from outside to be vented to burner to reach peak efficiencies. Proprietary parts pH has to be monitored on a regular schedule.
Compact size	
Factory assembled.	
Been installed in Europe for over 20 years.	

What Kind of Condensing Boiler?

There is a huge battle raging about which kind of boiler is the best type of condensing boiler. It boils down to two boiler types, stainless steel and cast aluminum. They both have pros and cons. In my opinion, I believe that they are both fairly comparable. The following is a comparison of aluminum and stainless steel boilers.

This is a comparison of Aluminum and Stainless Steel Boilers

Stainless Steel Boiler Compared to Aluminum Boilers	
Stainless Steel	
Advantages	**Disadvantages**
Stainless steel can handle acidic conditions found in the flue gases	Has lower heat transfer rate than aluminum
Aluminum	
Advantages	**Disadvantages**
Forms aluminum oxide to protect itself against corrosion. Can repair scratches that occur in the skin with the aluminum oxide. Transfer heat ten times faster than stainless steel. Has been used in automobile engines for over 25 years with good results.	Cannot use standard boiler cleaning soaps. Requires additive when using glycol.

Interesting Study

When choosing a boiler type, it is difficult to choose which type is right. I found an interesting study that was performed for a large school district in my area. The owner hired a national consulting engineering firm to perform a life cycle cost evaluation for boilers on three of their buildings. The three buildings were a high school, a middle school and an elementary school. The study compared several boiler types and manufacturers. The study used the US Department of Energy life cycle costing software to perform the study. I found the results interesting. Triad modular steel boilers had the lowest life cycle costs.

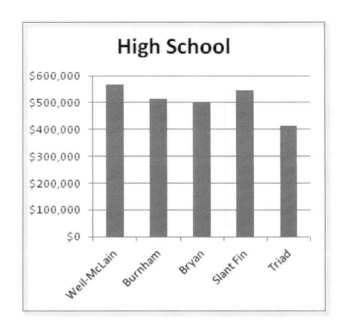

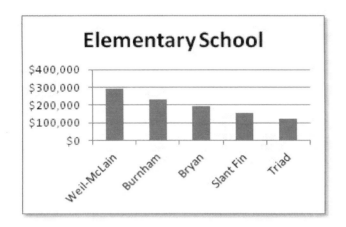

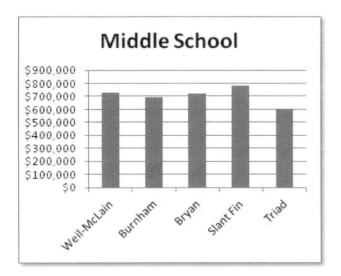

Chapter 14
Installing the Boilers

House Keeping Pad

Most boilers are installed on a concrete house-keeping pad. The pad is typically 4" or higher. It is used to keep dirt and water away from the boiler. If the building has a tendency to flood, you should increase the height of the pad. There 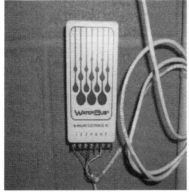 are also some switches that can be installed on the floor that will detect water and either alarm or shut off the boilers.

Planning the Location

When looking at the empty room, now is the time to think about how you want to set the boilers. Familiarize yourself with the boiler installation and God forbid, read the installation manual. Sometimes they are like trying to read a novel in a foreign language, but there are some key issues that are inside the manual. Some things to review: piping installation, burner service, fireside and waterside cleaning & access. Another reason to read the manual is to avoid a lawsuit. If the boilers are not installed as the manufacturer suggested, you may be the one sued when the boilers do not work.

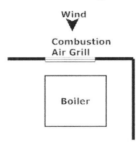

Location Note:

If you are installing a boiler with a pilot, be careful of installing it in a direct line of the combustion air louver. Wind could extinguish the pilot, resulting in a no heat call. On one project, the owner had to install a metal shield in front of the boiler to protect the burner from the wind.

Erection of Boiler

Some cast iron boilers require erection of the sections. These will be most likely very heavy. In addition, some boilers require special tools to assemble the boiler. If so, are the tools included in your price?

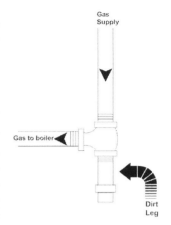

Gas Piping

The gas supply to the boiler should have a "Dirt Leg" to catch dirt and other items that might be in the gas line. The dirt leg should be 3" minimum in length. This is to protect the gas train components. Be careful when using Teflon tape or Teflon pipe dope. Some manufacturers will void the warranty if Teflon is used.

Block, Block & Bleed Gas Train Piping

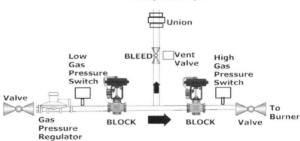

In some instances, you will need to install a Block, Block and Bleed gas train. Here are some pointers when installing this type of train.

✓ The bleed valve should be vented separately of the other gas train components.

✓ A union should be installed above the bleed valve to allow for testing of the bleed valve.

✓ Some solenoid bleed valves cannot be installed in a vertical position.

Gas Valve Leak Testing

It amazes me that we would never tolerate a leaking water valve on the sink but a gas valve is permitted a certain amount of leakage.

According to ASME CSD1, as well as most gas valve manufacturers, provisions should be made in the gas train to allow for leak testing of the electric safety shut off valves (SSOV). This is usually a ¼" tapping in the downstream nipple after the gas valve. CSD1 and the manufacturers recommend testing the valves at least once per year. If there are two electric SSOV's, then both have to be tested. The pilot solenoid valve should also be tested at least once per year. Some valves will have a downstream tapping as part of the valve. If the gas train contains a bleed valve, I would suggest testing this at the same time. This valve is piped to the outside and unless it is tested, a leak may never be found.

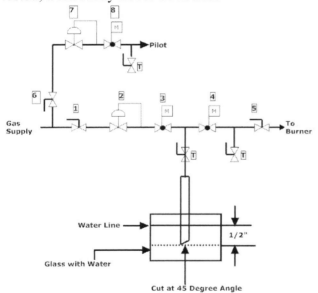

1	Upstream Manual Gas Valve
2	Gas Pressure Regulator
3	Safety Shutoff Valve
4	Safety Shutoff Valve or Blocking Valve
5	Downstream Manual Gas Valve
6	Manual Gas Valve- Pilot
7	Pilot Gas Pressure Regulator
8	Pilot Safety Shutoff Valve
T	Test Ports These may be part of valve

When testing the valves, refer to the manufacturer's recommendations. To meet the U.S. requirements, leakage must not exceed ANSI Z21.21, Section 2.4.2. It is based on air at standard conditions and limits leakage to a maximum of 235 cc/hr per inch of seal-off

diameter. This is not the same as pipe diameter. The following is the maximum bubble count for the valve sizes.

Valve Size (Inches)	Allowable Leakage (cc/hr)	Maximum Bubbles per 10 second test.
¾"	458	16
1'	458	16
1 ¼"	458	16
1 ½"	458	16
2"	752	26
2 ½"	752	26
3"	752	26
4" & Larger	1,003	35

Installing the Vent Piping

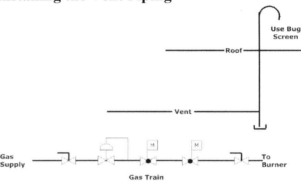

The vent from the gas train components may be gathered together into a common vent. When piping the vents together, a dirt leg should be installed to catch any rust or bugs that may have entered from the vent.

Vent Screens

When venting the gas train components outside, remember to install a bug screen over the outlet. This will prevent bugs from entering the vent tubing.

Piping the Boilers

What pipe material will you be using?
The most common pipe materials used on boilers are steel and copper. If you are using copper pipe, you need a dielectric fitting if it connects to another metal. A dielectric fitting should be installed at any point where there are two

dissimilar metals connected. If not, electrolysis will occur and leaks will develop.

Copper piping is not recommended for steam boilers.

Installing Hydronic Boilers

Using PVC, CPVC or Pex?

PVC or Polyvinyl Chloride is used throughout the HVAC industry. It can be used with a hydronic system but the temperature rating of the material limits its usage. PVC is rated for 140^0F. A mall in our area used PVC for their system. The heating system contained a boiler with water source heat pumps. The boiler control malfunctioned and overheated the piping, causing the PVC to melt. The piping looked like cooked spaghetti from each hanger.

CPVC or Chlorinated Polyvinyl Chloride is PVC that has been chlorinated. It has a higher temperature rating than PVC. CPVC is rated for 200^0F.

PEX is cross-linked polyethylene. It has been used since the 1960's. It has a higher temperature rating than PVC or CPVC. PEX has become widely used in our industry in applications such as radiant floor heat.

Temperature Ratings
PVC 140^0F
CPVC 200^0F
Pex 200^0F

Hydronic Water Feeder

It is a good idea to install bypass piping around the water feeder. When filling the system, this will save you some time. In addition, if the feeder ever fails, you could manually feed water into the loop.

Piping the header

When installing the new header, add some fittings for the thermometers and controls. At a minimum, you will need two on the return and two on the supply. When piping a Primary Secondary system, the supply and return to the boiler should be a maximum of 12" apart.

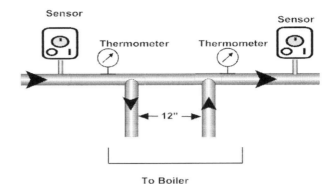

Piping Components

When piping a hydronic boiler, consideration should be given for the removal of air from the system. The air separator, or air removal fitting, removes the air inside the hydronic piping and diverts it to the expansion tank. Air inside a hydronic system can cause air pockets, reducing or completely shutting off the heat to certain portions of the building. This will result in comfort complaints.

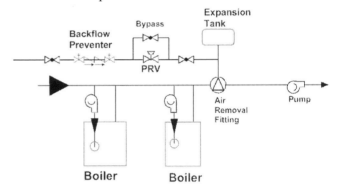

93

Installing the Steam Boilers

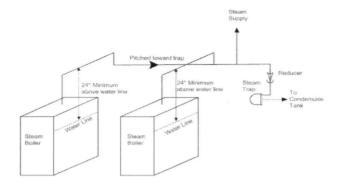

Piping the Steam Boiler

When piping the steam boiler, the steam header has to be a minimum of 24" above the water line. The reducer at the end of the steam main should be installed in the vertical piping. Maintain full pipe diameter until after the turn. This will reduce the chances of wet steam.
Remember the swing joint for a cast iron boiler. Calculate the steam velocity to the system. If it is too fast, you will have carry-over.

Steam Boiler Control Piping

Consider installing unions on pigtails or siphons so that the controls can be easily serviced or replaced.

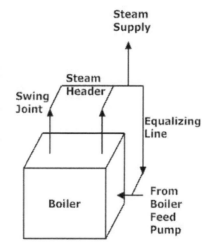

Misc Steam Items

When piping the low water cutoffs, consider using nipples and caps on the piping. This makes it easier to access the piping for cleaning. It is much easier to remove the piping with caps and nipples versus threaded plugs. The piping around the low water cutoff should be disassembled yearly and cleaned. The nipples going into the boiler can become plugged and render the low water cutoff inoperative. The discharge to the low water cutoffs should have a nipple installed that directs the discharge down.

How to calculate steam velocity

Lbs/Hr x Cubic Volume of Steam divided by 25 x Internal area of pipe = Steam Velocity

$$\text{Steam Velocity} = \frac{Lbs\ Hour \times Cubic\ Volume\ of\ Steam}{25 \times Internal\ Area\ of\ Pipe}$$

Lbs. Steam/Hr = Btuh/960
Lbs. Steam/Hr = Boiler HP x 34.5

Volume of Steam in Cubic Feet per Hour	
PSIG Steam	Cubic Feet per pound
0	27
1	25
2	24
3	22
4	21
5	20

Internal Volume of Schedule 40 Pipe	
Pipe Size	Internal Square Inches
2"	3.36"
2 ½"	4.78"
3"	7.39"
4"	12.73"
5"	19.99"
6"	28.89"
8"	51.15"
10"	81.55"
12"	114.80"

Gauge Glasses

When installing a new steam boiler, you will most likely have to cut the gauge glass that is shipped with the new boiler. They usually come in different lengths such as 12", 18" and 24". There is a special tool called a Gauge Glass Cutter that makes cutting the gauge glasses easier. They can be purchased at the same place that you purchased the boiler. These are a time saver when installing a steam boiler. If you do not have one, a file can be used to trim the gauge glass. It is time consuming and difficult to do. To use a file, you have to keep running the file over the line that marks the desired length of the gauge glass. After the glass is etched, the gauge glass can be tapped to break it. In most instances, the gauge glass does not break evenly. I would recommend an extra gauge glass, brass washers and rubber gaskets when you are installing a steam boiler. The gauge glasses have a habit of breaking on the job site. When installing the gauge glass onto the boiler, slide the brass nut onto the gauge glass. Then slide the brass washer followed by the rubber gasket. The brass washer and rubber gasket should fit inside the brass nut. When tightening the nuts onto the fittings, they should only be tightened until they do not leak. Do not over-tighten as they will break the glass.

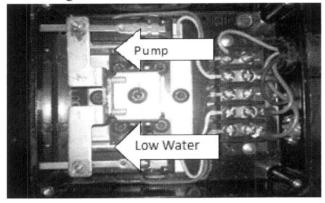

The picture shows the inside of a McDonnell Miller #150 combination low water cutoff and pump control. One internal switch will start or stop the boiler feed pump. The other will shut off the burner if the water level drops too low.

Servicing the Low Water Cutoff

The low water cutoff should be serviced on a regular basis. When servicing these, you should purchase a gasket that will be used to re-attach the head assembly. Typically, you will disassemble these yearly. The inside of the bowl will require cleaning so that the float moves freely inside the chamber. The new low water cutoffs use a probe instead. When reassembling the head of the low water cutoff, both sides have to be cleaned and free of grit or the gasket will not seal. The piping connections to the boiler should be inspected and cleaned.

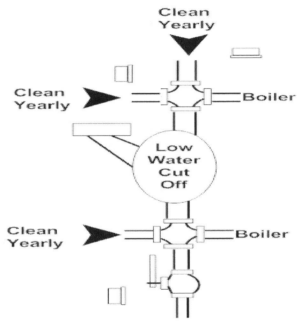

When piping the low water cutoff, you should consider using pipe caps and nipples instead of pipe plugs. It will make it much easier to access the pipe connections to the boiler.

When I start up the boiler, I will mark the normal water levels on the gauge glass. This makes it easier when trouble shooting in the future.

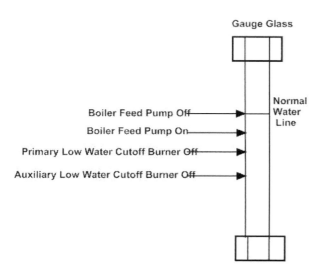

Gauge Glass

Normal Water Line

Boiler Feed Pump Off

Boiler Feed Pump On

Primary Low Water Cutoff Burner Off

Auxiliary Low Water Cutoff Burner Off

Did You Know…
A pound of 180^0f condensate contains 148 Btu.

Did You Know…
A 1,000,000 Btu boiler will require 206 CFM air for the burner?
Based upon 30% excess air

Chapter 15
Boiler Venting

Venting the new boilers
Because of the complexity of flue design, this book will not cover boiler flue design. I would suggest that you contact the manufacturer or a flue manufacturer's representative in your area. What worked on a previous job may not work on this project. Each project is different.

Traditional Venting
Before reusing the existing flue, verify that it can be used with the new boilers. If the boilers are over 83% efficient, they may not be able to be vented into the existing flue. As a boiler approaches 84% efficient, the flue gases could condense and destroy the breeching and chimney. This is especially true if the chimney is on an outside wall.

Draft Hood Appliances
Hot gases rise and draft upward into the chimney. The draft hood allows dilution air to vent and mix with the flue gas, which reduces the humidity or dew point and thus reduces chances of corrosion. The dew point in a gas vent is about $90^{\circ}F$ to $130^{\circ}F$.

Boiler Venting Types
There are four basic types of boiler vents. The following is a description of each. The boiler manufacturer will inform you as to what type of venting they require.

Category 1	Appliances that operate with zero or negative pressure at the appliance outlet and at temperatures where condensation of flue gases IS NOT likely. AFUE 65% - 83%, can be vented with single wall or Type B vent. This is the most common type boiler.
Category 2	Appliances that operate with 0 or negative pressure at the appliance outlet and at temperatures where condensation of flue gases IS likely. This is very rare with the newer boilers, and these must be vented with corrosion-resistant materials like stainless steel.
Category 3	Appliances that operate with positive pressure at the appliance outlet and at temperatures where condensation of flue gases IS NOT likely. AFUE 78% - 83%, These are vented with airtight high temperature plastic or airtight single wall metal.
Category 4	Appliances that operate with positive pressure at the appliance outlet and at temperatures where condensation of flue gases IS likely. AFUE 90%+, These are vented with airtight high temperature plastic, stainless steel or CPVC.

Draft Control
If your Category 1 type boiler has a stack in excess of 30 feet tall, draft controls should be considered. The most common draft controls in commercial buildings are barometric dampers.

Barometric Damper
Barometric dampers are used on Category 1 type boilers. A barometric damper is device that helps to regulate the draft in a stack or chimney of a heating system by opening and introducing boiler room air into the stack when the draft is excessive. Excessive draft can cause numerous problems including higher fuel costs, flame impingement and elevated carbon monoxide

levels. The following is a description of how a barometric damper operates.

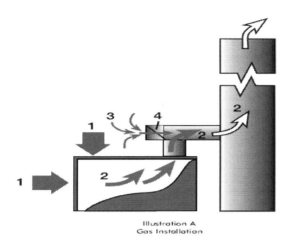

Illustration A
Gas Installation

Static pressure of the cool air (1) Illustration A exerts pressure on the outside of the furnace or boiler, the breaching, and stack. The pressure difference between the room air and heated gas (air) causes products of combustion (2) to flow (draft) through the unit and rise through the breaching and chimney. Room temperature air (3) enters through the barometric draft control (4) in the precise amount needed to overcome the excess drafts caused by temperature variations, wind fluctuations, and barometric pressure changes. Combustion of fuel is complete and the process is stabilized. The velocity of combustion gases through the heat exchanger is slowed so more heat is extracted. The unit operates more efficiently, reliably, and requires less maintenance." Courtesy of Fields Controls

Choosing a barometric damper

Barometric Damper

When using a barometric damper on your boiler, you need to choose which type. If the boiler is fired with only natural gas, you need a double acting barometric damper. The double acting barometric damper will open and could allow flue gases to spill into the boiler room if there is a blockage in the flue. You

should also use a spill switch that will shut off the fuel if spillage is sensed. If you are firing with #2 fuel oil, you will need one with a stop.

Installing the Barometric Damper

The following are the preferred locations for the installation of barometric dampers according to Fields Controls. A is the best location. On some boilers that have turbulence in the flue, a location may cause spillage.

In most instance, the barometric damper should be the same size as the flue in which it is installed.

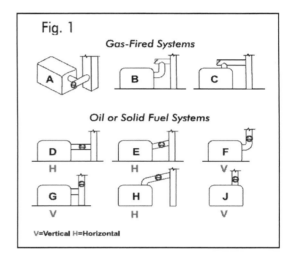

A = Best Location
B = Better Location
C = Good Location

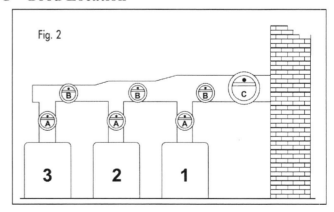

98

How Does Draft Affect the Flame?

What is a decimal point among friends? The frantic call came into the office. The installer asked for someone to come right away to his jobsite. He said that the steam boiler water level was bouncing and he was even getting water pouring down the top of the gauge glass. When we arrived at the jobsite, we noticed the exact phenomenon that he described. I visually inspected the flame and saw that it was being drawn into the boiler. I asked if he had checked the boiler draft. The boiler was designed for about -0.05" w.c. The installer assured me that it was. When we checked the draft, it was -0.50" w.c. I explained that the draft was too high and he needed a draft control. "What's the difference?" he asked. I explained that his current draft was ten times more than the boiler was designed for. Excessive draft can cause several problems. The first is impingement. When flames hit the colder metal, it can cause damage to the boiler. In addition, it will cause carbon monoxide to form in the flue gases. It could also wreak havoc on the boiler water levels. As in this job, the flames were drawn into the rear tubes heating the rear of the boiler more than the front. This caused the water in the rear of the boiler to rise and fall forward. This is what was causing the water to pour down the front of the gauge glass. The second problem with excessive draft is that it causes the burner to over fire. The gas pressure regulator adjusts the gas pressure to the outlet pressure. If the boiler is excessively negative, the pressure regulator will open more trying to increase pressure. This will over fire the boiler.

Sidewall venting

Many new boilers can be sidewall vented. That leads to some interesting challenges. If you read the codes for sidewall venting, I dare you to meet the codes when installing a sidewall vented boiler in an older building

- "Where adjacent to walkways, the termination of mechanical draft systems shall not be less than 7 feet above walkway"
- "3 feet above any forced air inlet within 10 feet

- 4 feet below, 4 feet horizontally from or 1 foot above any door, window or gravity vent into building
- No closer than 3 feet from an interior corner formed by 2 walls perpendicular to each other
- Not within 3 feet horizontally or directly above an oil tank or gas meter
- At least 12 inches above finished grade

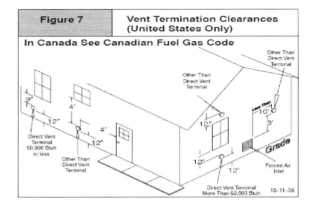

A Caution about Sidewall Venting

Sidewall Vent

A nursing home in my area had a new heating system installed that used sidewall venting. The boiler was vented through the wall into an area where two perpendicular walls met. The engineer and contractor installed the venting as per the code. The system worked well the first heating season. During the second season, a dangerous condition arose. During a cool rainy day, the combustion gases would not vent due to the weather conditions. Some of the flue gases gathered in the ell formed by the two perpendicular walls and was drawn into the ventilation inlets of the nursing home unit ventilators. The patients had to be evacuated. A lawsuit was filed and the engineer and contractor had to install a stainless steel vent that rose above the four-story structure at their own cost.

A Note on venting Category 3 and Category 4 Boilers.

When venting these boilers, use caution when using a common breeching. Since each boiler features a positive pressure in the flue, this could force the flue gases out of the idle boiler and damage the boiler. In addition, it could allow flue gases to flow out of the idle boilers and into the boiler room.

Width	Circular Equivalent of Rectangular Duct					
	Height in Inches					
	6	8	10	12	14	16
6	7	8	8	9	10	10
8	8	9	10	11	11	12
10	8	10	11	12	13	14
12	9	11	12	13	14	15
14	10	11	13	14	15	16
16	10	12	14	15	16	17
18	11	13	15	16	17	19
20	11	13	15	17	18	20
22	12	14	16	18	19	20
24	12	15	17	18	20	21
26	13	15	17	19	21	22
28	13	16	18	20	21	23
30	14	16	18	20	22	24
36	15	17	20	22	24	26
42	16	19	21	23	26	28
48	17	20	22	25	27	29
54	17	21	23	26	29	31
60	18	21	24	27	30	32

Width	Circular Equivalent of Rectangular Duct					
	Height in Inches					
	18	20	22	24	26	28
6	11	11	12	12	13	13
8	13	13	14	15	15	16
10	15	15	16	17	17	18
12	16	17	18	18	19	20
14	17	18	19	20	21	21
16	19	20	20	21	22	23
18	20	21	23	23	24	24
20	21	22	24	24	25	26
22	22	23	25	25	26	27
24	23	24	26	26	27	28
26	24	25	27	27	28	29
28	24	26	28	28	29	31
30	25	27	31	29	31	32
36	27	29	33	32	33	35
42	29	31	35	34	36	37
48	31	33	37	37	38	40
54	33	35	39	39	40	42
60	34	37	39	40	42	44

Width	Circular Equivalent of Rectangular Duct					
	Height in Inches					
	30	36	42	48	54	60
6	14	15	16	17	17	18
8	16	15	19	20	21	21
10	18	17	21	22	23	24
12	20	20	23	25	26	27
14	22	22	26	27	29	30
16	24	24	28	29	31	32
18	26	26	29	31	33	34
20	27	27	31	33	35	37
22	28	29	33	35	37	39
24	29	31	34	37	39	40
26	31	32	36	38	40	42
28	32	35	37	40	42	44
30	33	36	39	41	4	46
36	36	39	42	45	48	50
42	39	42	46	49	52	55
48	41	45	49	52	56	59
54	44	48	52	56	59	62
60	46	50	55	59	62	66

Chapter 16
Boiler Installation Codes

When designing and installing a replacement boiler, there are several code books and many codes. The following are some pertinent excerpts about boilers from the code books. Every contractor should have a copy of these codes.

The International Building Code (IBC) is adopted at the state or local level in 50 states plus Washington, D.C.

The International Residential Code (IRC) is adopted at the state or local level in 48 states plus Washington DC, U.S. Virgin Islands

The International Fire Code (IFC) is adopted at the state or local level in 42 states plus Washington, D.C.

The International Energy Conservation Code (IECC) is adopted at the state or local level in 42 states plus Washington, D.C.

The International Plumbing Code (IPC) is adopted at the state or local level in 34 states plus Washington DC, Puerto Rico

The International Mechanical Code (IMC) is adopted at the state or local level in 47 states plus Washington, D.C.

The International Fuel Gas Code (IFGC) is adopted at the state or local level in 43 states plus Washington, D.C.

In other words, it is coming to a neighborhood near you. I would recommend purchasing the set of codes from the International Code Council.
The following information was provided by International Code Council.

International Energy Conservation Code 2006

Excerpted for the 2006 International Energy Conservation Code, copyright 2006
Washington, DC: International Code Council
Reproduced with permission
302.1 Interior design conditions. The interior design temperatures used for heating and cooling load calculations shall be a maximum of 72 degrees F for heating and a minimum of 75 degrees F for cooling.

503.2.1 Calculation of heating and cooling loads. Design loads shall be determined in accordance with the procedures described in the ASHRAE Fundamentals Handbook. Heating and cooling loads shall be adjusted to account for load reductions that are achieved when energy recovery systems are utilized in the HVAC system in accordance with the ASHRAE HVAC Systems and Equipment Handbook. Alternatively, design loads shall be determined by an approved equivalent computation procedure, using the design parameters specified in Chapter 3.

503.2.2 Equipment and system sizing. Heating and cooling equipment and systems capacity shall not exceed the loads calculated in accordance with Section 503.2.1. A single piece of equipment providing both heating and cooling must satisfy this provision for one function with the capacity for the other function as small as possible, within the available equipment options.

Exceptions:
- Required standby equipment and systems provided with controls and devices that allow such systems or equipment to operate automatically only when the primary equipment is not operating.

- Multiple units of the same equipment type with combined equipment type with combined capacities exceeding the design load and provided with controls that have the capability to sequence the operation of each unit based on the load.

503.2.8 Piping insulation. All piping serving as part of a heating or cooling system shall be thermally insulated in accordance with Table 503.2.8

Exceptions:

- Factory-installed piping within the HVAC equipment tested and rated in accordance with a test procedure referenced by this code.

- Piping that conveys fluids that have a design operating temperature range between 55 degrees F and 105 degrees F.

- Piping that conveys fluids that have not been heated or cooled through the use of fossil fuels or electric power.

- Run out piping not exceeding 4 feet in length and 1" in diameter between the control valve and the HVAC coil.

Minimum Piping Insulation

Fluid	Pipe 1 ½" and below	Pipe Larger than 1 ½"
Steam	1 ½"	3"
Hot Water	1"	2"
Chilled water, brine or refrigerant	1"	1 ½"

Minimum Efficiencies
Boiler, Gas Fired

Capacity	Hot Water	Steam
<300,000 Btuh	80% AFUE	75% AFUE
300,000 – 2,500,000 Btuh	75%	75%
>2,500,000 Btuh	80%	80%

Boiler, Oil Fired

Capacity	Hot Water	Steam
< 300,000	80% Afue	80%
300,000 – 2,500,000 Btuh	78%	78%
>2,500,000 Btuh	83%	83%

503.2.9.3 Manuals. The construction documents shall require that an operating and maintenance manual be provided to the building owner by the mechanical contractor, the manual shall include, at least, the following:

1. Equipment capacity (input and output) and required maintenance actions.
2. Equipment operation and maintenance manuals.
3. HVAC system control maintenance and calibration information, including wiring diagrams, schematics, and control sequence descriptions. Desired or field-determined set points shall be permanently recorded on control drawings, at control devices or, for digital control systems, in programming comments
4. A complete written narrative of how each system is intended to operate.

503.4.3 Hydronic systems controls. Hydronic heating systems comprised of multiple-packaged boilers and designed to deliver conditioned water or steam into a common distributions system shall include automatic controls capable of sequencing operation of the boilers. Hydronic heating systems comprised of a single boiler and greater than 500,000 Btuh input design capacity shall include either a multistage or modulating burner.

503.4.3.4 Part load controls. Hydronic systems greater than or equal to 300,000 Btuh in design output capacity supplying heated or chilled water to comfort conditioning systems shall include controls that have the capability to:

1. Automatically reset the supply water temperatures using zone-return water temperature, building-return water temperature, or outside air temperature as

an indicator of building heating or cooling demand. The temperature shall be capable of being reset by at least 25% of the design supply to return water temperature difference; or

2. Reduce system pump flow by at least 50% of the design flow rate utilizing adjustable speed drive(s) on pump(s), or multiple staged pumps where at least ½ of the total pump horsepower is capable of being automatically turned off or control valves designed to modulate or step down, and close, as a function of load, or other approved means.

503.4.3.5 Pump isolation. Boiler plants including more than one boiler shall have the capability to reduce flow automatically through the boiler plant when a boiler is shut down. Please see the Primary Secondary system that illustrates this below.

International Plumbing Code 2006
Excerpted for the 2006 International Plumbing Code, copyright 2006
Washington, DC: International Code Council
Reproduced with permission

608.16.2 Connections to boilers. The potable water supply to the boiler shall be equipped with a backflow preventer with an intermediate atmospheric vent complying with ASSE 1012 or CSA B64.3. Where conditioning chemicals are introduced into the system, the potable water connection shall be protected by an air gap or a reduced pressure principle backflow preventer, complying with ASSE 1013, CSA B64 or AWWA C511.

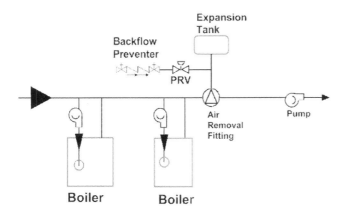

401.4.1 Intake openings. Mechanical and gravity outdoor intake openings shall be located a minimum of 10 feet (3048 mm) horizontally from any hazardous or noxious contaminant source, such as vents, chimneys, plumbing vents, streets, alleys, parking lots and loading docks, except as otherwise specified in the code. Where a source of contaminant is located within 10 feet horizontally of an intake opening, such opening shall be located a minimum of 2 feet below the contaminant source.

The exhaust from a bathroom or kitchen in a residential dwelling shall not be considered to be a hazardous or noxious contaminant.

Table 401.5 Opening size in louvers, grilles and screens

Outdoor Opening Type	Minimum & Maximum sizes in louvers, grilles and screens
Exhaust Opening	Not < ¼" and not > ½"
Intake Opening in residential occupancies	Not < ¼" and not > ½"
Intake Opening in other than residential occupancies	Not < ¼" and not > ½"

1004.7 Operating adjustments and instructions. Hot water and steam boilers shall have all operating and safety controls set and operationally tested by the installing contractor and boiler operating instructions shall be furnished by the installer for each installation.

1005.1 Valves. Each boiler and modular boiler shall have a shutoff valve in the supply and return piping. For multiple boiler or multiple boiler installations, each boiler or modular boiler shall have individual shutoff valves in the supply and return piping.

Exception: Shutoff valves are not required in a system having a single low pressure steam boiler.

1006.5 Installation. Valves shall not be located on either side of a safety or relief valve connection.

1006.6 Safety and relief valve discharge. The discharge pipe shall be the same diameter as the safety or relief valve discharge.

1007.1 All steam and hot water boilers shall have be protected with a low water cut off control

1008.1 Every steam boiler shall be equipped with a quick-opening blow-off valve.

Chapter 17
Cleaning the Boiler

Steel or Cast Iron Hydronic Boilers

Consult with the manufacturer about the products they recommend. In most instances, Tri Sodium Phosphate or TSP is mixed with the water and circulated for 2-3 hours. Drain, flush and refill with fresh water. After flushing the cleaner from the system, a pH test should be done to verify that the water is within the proper range. The typical pH range is 7.0 - 8.5. Please check with the manufacturer to verify their requirements. Some cast iron boilers have special requirements due to the material that is used between the sections.

One boiler manual that I saw has the following instructions for cleaning a new steam boiler.
Use equal parts of
Trisodium Phosphate
Caustic Soda
Soda Ash
It recommends using 1/2 pound per boiler horsepower. It suggests operating the steam boiler, without allowing it to steam for 16-24 hours. It also suggests allowing the condensate to be wasted to the drain for 7-10 days and checking the water treatment chemical levels every four hours. See, you really need to read the owner's manual prior to providing a firm price. This startup lasted for days.

> **Note: If you have an aluminum water boiler, you may not be able to use Tri Sodium Phosphate (TSP) as a cleaning agent for the boiler. TSP has a high pH level and could remove the natural protective oxide layer from the aluminum.**

Steel or Cast Iron Steam Boilers

When cleaning a steam boiler, do not allow the TSP to boil off into the piping. A typical application is to heat the boiler to 180 degrees F and allow it to stay there for 2 hours. After, drain, flush and refill the system. Check the pH level of the water to make sure it is within the manufacturers recommended limits.

I like to let the condensate run to the drain for a couple days when it is started to allow the system to purge itself of the contaminants.

Skimming the Boilers
Steam Boilers
Skim Fitting

Most steam boilers have a skim fitting which is located near the water line so that the solids can be skimmed from the boiler. This will help to keep the boilers running efficiently.

Did You Know…
The Titanic had 29
boilers and 159 furnaces

Did You Know…
Eniac, thought of as the
world's first modern
computer, was built in
1944. It measured 8 feet
x 3 feet x 100 feet. It
could process 5,000
basic arithmetic
operations per second. It
weighed over 60,000
pounds.

The Intel core2
computer chip,
introduced in 2008, can
perform 54,455,000,000
calculations per second.
It weighs less than a
packet of sweetener.

Chapter 18
Glycol for the System

If you are contemplating the use of glycol in your hydronic system, there are several important factors to consider.

Can Your Boiler Tolerate Glycol?
Before installing glycol in your system, check with the boiler manufacturer. Some aluminum boilers do not react well with glycol. Glycol designed for aluminum boilers can be ordered. Some boiler manufacturers have specific requirements when adding glycol for their boilers.

Reduced Efficiency
Adding glycol to your system will decrease the heating capacity of the system. For example, when you add 20% propylene glycol to the system, your heating capacity drops to 97% of a system without glycol. At 50% concentration, your capacity drops to 90%. Your new 80% boiler suddenly dropped to a little over 70% efficient. This should be factored in when sizing the new system.

Expansion Tank Sizing
If you are using glycol in the system, the expansion tank should be 20% larger than one for only water.

Reduced Pump Capacity
You may need to increase the size of the pump in your system due to the lower heating capacity. If you have a 50% mixture, you will need 110% of your normal pumping capacity

Increased Maintenance
Systems containing glycol require extra maintenance. The pH has to be checked yearly. The glycol composition has to be checked using a refractometer. Glycol will breakdown eventually and may cause damage to your system. If the concentration level drops in your tests, this should indicate a piping leak. The leak should be found and repaired.

Cleaning the system
The system should be cleaned prior to introduction of glycol. Flush the system with a heated 1-2% solution of Trisodium phosphate for 2 to 4 hours, then drain and rinse thoroughly. This will remove excess pipe dope, cutting oils and solder flux.

Automotive Glycol?
This is not recommended as it was not designed to be used in a hydronic system. Automotive antifreeze is formulated with silicates, which tend to gel, reducing heat transfer efficiency. Use an inhibited glycol designed for heat transfer applications.

Which Type of Glycol?
You can choose between ethylene and inhibited propylene glycol. Uninhibited glycol is very corrosive and could lead to damage in your system. Ethylene glycol is toxic to humans and animals. A special permit may be required when using it. Ethylene glycol will provide slightly better freeze protection than propylene glycol. Propylene glycol is more environmentally friendly and not toxic.

How Much Glycol?
The most common percentage is between 20% to 50% concentration of glycol.

Water Quality
Water chemistry is a concern when you introduce glycol into your system. Poor water quality can lead to scale, sediment deposits, or the creation of sludge in the heat exchanger which will reduce heat transfer efficiency. It can damage the system by depleting the corrosion inhibitor and promoting a number of corrosions including general and acidic attack corrosion. Before using the local water, it should be analyzed by an water treatment expert. Good quality water contains:

- Less than 50 ppm of calcium
- Less than 50 ppm of magnesium
- Less than 100 ppm (5 grains) of total hardness
- Less than 25 ppm of chloride

- Less than 25 ppm of sulfate

The glycol should be mixed at room temperature with either demineralized or deionized water if the water quality is questionable.

The safest, although not the least expensive, solution may be to order premixed glycol from the manufacturer.

Freezing Point

Concentration by volume	Ethylene Glycol	Propylene Glycol
50%	-37F	-28F
40%	-14F	-13F
30%	+2F	+4F
20%	+15F	+17F

Maintenance

The glycol must be checked at least once a year in accordance with the manufacturer's recommendations. A base line analysis should be performed within two to four weeks of initial mixing. This measurement will be used to verify that the fill was completed properly, and will serve as a reference point for comparison with future test results. As a bare minimum, the solution should be analyzed for glycol concentration, solution pH and general fluid quality.

Solution Testing

If you are using a 30% to 50% solution, the pH should be between 8.3 and 9.0. If the level falls below 8.0, this could indicate lowered inhibitors. In some instances, inhibitors can be added. If the pH falls below 7.0, the glycol should be removed and flushed. This level of pH could cause damage to the boiler and piping.

Mark the System

When you are done with the system and turn it over to the owner, the system should have clear signage that tells anyone working on the system that it contains glycol. It should also have the following information:

- Type of glycol that is used in the system, ethylene or propylene
- Concentration of glycol
- Initial pH readings
- Installation date of glycol
- Last test date

Some Glycol Considerations

Galvanized pipe should not be used in the hydronic system as the Zinc could have an adverse effect and form sludge

Make sure the system is clean before filling. Pre-fill flushing is highly recommended.

Mix the solution at room temperature.

In order to minimize the possibility of glycol loss due to undetected leaks, hydrostatically test the system for 24 hours prior to filling.

Never use a chromate water treatment in a system with glycol. The chromate will damage the glycol and can lead to severe system degradation.

Do not use in a system that may have a solution temperature over 300F.

Do not use check valves or closed zone valves that would isolate a part of the system, preventing proper expansion and resulting in freeze damage.

A strainer, sediment trap, or some other means for cleaning the piping system must be provided. It should be located in the return line ahead of the boiler and pump. This must be cleaned frequently during initial operation.

Automatic make-up water systems should be avoided in order to prevent undetected dilution or loss of glycol.

Check local codes to see if systems containing these solutions must include a back-flow preventer, or an actual disconnect from city water lines.

Do not use glycol in steam systems.

Did You Know…
Hydronic systems
expand at 3% of its
volume when heated.

Did You Know…
Did you know that it
takes about 1,220 Btu to
change one pound of
snow to water vapor?

Chapter 19
Boiler Start Up

The start-up of the boiler is very important. The equipment manufacturer will usually supply a checklist that they want performed during the start-up. Some of the tools you will need for a startup are:

- Combustion analyzer
- Manometers
- Voltmeter
- Amp Clamp Meter
- Screw Driver
- Operations and Maintenance Manual

Starting the New Boilers
Training Session
When starting the boilers, it is a good idea to give a short training session for the new owner. In the instructions, show them the safety controls and normal maintenance. Some people will record the training session. This is a good idea as they could refer to it at the beginning of the next heating season or if they hire a new person.

Boiler Start-Up
Items that should be checked on your start up:
Manual gas valve
Electric gas valve
Gas Pressure Switch
Pilot Assembly
Igniter
Ignition Transformer
Flame Signal Pilot and Main Flame
Flame Scanner
Gas Pressure Regulator
Gas Train Venting
Bearings & Linkages
Pressure Reducing Valve Water Feeder
Modulating Motor
Modulating Control
Low Fire Start Switch
Flame Safeguard
Thermocouple
Operating Controls
Limit Controls Manual Reset
Air fuel interlock
Low water Cut-off / Pump Control

Auxiliary Low Water Cut-off Manual Reset
Flow switch on copper boilers
Water Column
Main Burner
Burner Mounting Plate
Refractory
Blower wheel
Blower motor
Wiring Connections
Gauges/Thermometers
Draft Controls
ID Fan The Id fan is an induced draft fan that is installed in the flue. It will pull the flue gases from the boiler.
Flue
Piping
Expansion Tank
Relief Valve and Piping

All Boilers
Prior to Start-Up
Pictures
In this world of litigation, protect yourself by taking pictures of the new boiler and piping.

Wiring Connections
If you want to save yourself a future service call, check all the wiring connections for the burner and controls. A loose wiring connection could cause some strange problems.

Gas Pressure
Prior to starting the boilers, the incoming gas pressure up stream of the regulator should be measured to see if it is within a tolerable range. It is always a good idea to keep the upstream manual gas valve that is located before the gas pressure regulator and gas valves closed until you verify that the gas pressure is within the desired range.

Flue
Verify that the flue is installed as per the manufacturer's recommendations. Check all joints and note the size and pitch of the flue.

Piping

Verify that the piping is installed as per the manufacturer's recommendations. Note size of piping.

Combustion Air

Verify that there is adequate combustion air for the room. Add up all the fuel-fired equipment including the water heater. If there are motorized combustion air dampers, make sure that they work properly. If using motorized dampers, an end switch should be used to verify that the dampers are fully open prior to the burner starting. When using a make-up air fan, the operation of the fan should be verified before the burner is started.

Gas Train Piping

Verify that all components are installed correctly. This would include venting of the gas train components that require venting. If you use Teflon Tape on the gas train components, be sure that the manufacturer permits it.

Blower Motor

Verify that the blower motor rotation is correct. This should be done prior to any fuel being turned on. Verify that the wheel is aligned properly and not loose on the shaft.

Venting the Hydronic System

Venting a hydronic system could add hours to your start up. Sometimes, you may have trapped air pockets in the hydronic system. A way to free up the air pockets may be to start and stop the pump. This may break the air pocket free.

Operating & Limit Controls

Verify that all operating and limit controls are installed correctly. In most instances, the system will need both an operating control and a limit control. The operating control will be set for the intended operating temperature or pressure. The limit control will be set higher. The limit control would most likely be a manual reset type control.

On some installations, like for the federal government, there are tougher regulations. For example, at one boiler project for the government, the inspector wanted us to "jumper" the operating and safety controls to verify that the relief valve operated. The boiler runs until the relief valve opens. On this particular project, it was a high-pressure boiler. If you have never witnessed this, it is intimidating.

The controls should all be tested and verify that they operate. Some techs will manually lower the setting on the control to see if it shuts off the burner. I prefer setting it at the desired set point and watch how close the set point is to the actual shutoff point. This should be noted in the job file.

Never try repairing the safety controls, which include the operating, limit, low water cut-off or flame safeguard controls. You could void the warranty and make yourself personally responsible for any problems.

Controlling our New Boiler
How low can we set the boiler?

Check with the manufacturer to see what they recommend. Most traditional boilers will condense if the water temperature is below 140^0 F on the return.

Boiler Sequencing

What is the best way to sequence boilers? I was discussing the boiler sequencing on a new project with the "Out of town" control expert that the client had hired.

"I always change the lead boiler each time there is a call for heat. In that way, we even the run time of all the boilers." He said with a condescending smile.

I respectfully disagreed with his logic. I explained that with boilers piped in a primary secondary arrangement, you should consider firing the same lead boiler all the time for certain amount of time. When you fire a different boiler each time, you are firing into a cold boiler since they are isolated. This leads to flue gas condensation and erosion of the fire side of the boilers. If you are firing into a boiler that is already warm, the flue gases will be less likely to condense. This will extend the life of the system. I prefer to have one boiler as the lead boiler for a month or two. After that, I would change the lead boiler to the next boiler and cycle the previous lead boiler as the last boiler to fire.

Another consideration of boiler sequencing is the boiler arrangement and their proximity to the chimney. In the picture below, you will see that this project has five boilers. If I were controlling this project, I would have Boiler #5 as my lead boiler. This will preheat the flue and chimney. If Boiler #1 was the lead boiler, we could have the flue gases condense before they reach the chimney.

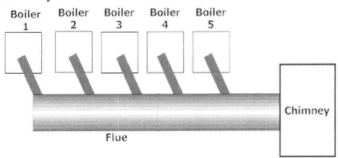

Relief Valve
The relief valve should be installed as per manufacturer's recommendations. Note the relief valve pressure setting. The piping from the boiler to the relief valve and the discharge piping from the relief valve should not have a valve. The discharge pipe has to be the same diameter as the relief valve discharge pipe size. The discharge piping should be supported so that the weight of the piping is not on the relief valve. The drain on the relief valve should be piped close to the ground. It should be left unthreaded or cut on an angle so that no one will install a cap in the event of leakage. If the boiler is a steam boiler rated at 500,000 of higher, the relief valve must be vented outside with a drip pan ell.

System Pressure
One pound of pressure will lift water 2.3 feet. If the highest radiator is 100 feet high above the boiler room, you will need to multiply the height of the highest riser by 0.43. In addition, you will need an additional 4 pounds of pressure as a safety factor, according to Bell & Gossett. Our building will require the following pressure (100 feet x 0.43) = 43 Psi + 4 Psi = 47 Pounds of pressure. A quick rule of thumb to see how much pressure we need for a tall building is to divide the elevation of the tallest radiator by 2.

Pilot Assembly

Ignition Electrode
Prior to start-up, the pilot assembly should be pulled and inspected. Verify the ignition electrode is not cracked and is adjusted to the proper settings. Be careful not to drop the electrode as it will have porcelain covering over the electrode and is very fragile. Check for cracks in the porcelain. The ignition electrode is connected to a transformer. The transformer will provide between 6,000 and 10,000 volts to the electrode.

Ignition Transformer

Burner Mounting Plate
Verify the mounting plate is installed correctly with no gaps or leaks. It should be sealed with a gasket or rope so that no products of combustion could leak.

Air Venting
A good technician that I know opens a bleed valve on the piping while starting the steam boiler until the steam is venting. This eliminates much of the air in the system.

Cold Starting

When starting a cold steam boiler, always keep the burner at low fire until it starts to steam then allow it to go to high fire.

Rebuilt Flame Safeguard Controls

Be careful buying rebuilt flame safeguard controls, as they may not meet UL approval. According to Honeywell, the UL label is void if another firm rebuilds them. I would not want to take that chance just to save a few dollars.

Flame Safeguard

One the most common questions that I am asked is why they use flame safeguard controls when an inexpensive thermocouple does the same thing. They, indeed do the same job; shutting off the burner in the event that no flame is sensed. The only difference is response time. A thermocouple may take between one to three minutes to respond. A modern flame safeguard will react within 2-4 seconds. This could be a huge difference. For example, if a 120,000 Btuh boiler equipped with a thermocouple sensed a flame loss and shut off the burner after a minute, the boiler would allow 2 cubic feet of unburned gas into the boiler. When combined with the combustion air, there would be about 24-30 cubic feet of combustible material in the boiler. Now, what would happen if we had the same thermocouple on a boiler rated for 1,200,000 Btuh? If the thermocouple reacted in one minute to a loss of flame, there would be 20 cubic feet of gas and about 240 - 300 cubic feet of combustible material. This could create a big bang in the boiler room.

Annunciator

If you service boilers on a regular basis, the annunciator for the Honeywell flame safeguards in the picture is great service tool. It will plug into the Honeywell 7800 series flame safeguards. It allows you to pull up the fault history of the flame safeguard as well as displaying the flame signal.

It's All in How You Say it?

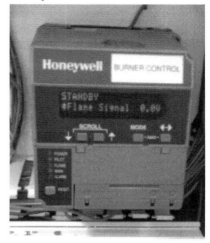

I was asked to consult on a boiler project in Atlanta and took my combustion analyzer with me through the airport. When the person at the airline counter asked what was in the package, I told them it was a combustion analyzer. I was immediately escorted to a small room with a metal desk. Two men with white shirts and ties asked me a series of questions. They thought I was a red headed terrorist. After twenty minutes, I was able to convince them that I was there to service some boilers. On the way home, when asked what I was carrying on, I said it is a Boiler Efficiency Tester and I was sent

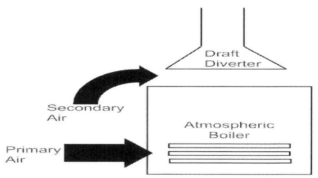

right through.

Combustion analysis

Let us pretend that we have invented the perfect burner. It is 100% efficient. To get perfect (100%) efficiency, our burner needs 10 parts of air to 1 part of gas. The only problem is that our perfect burner will require constant attention. If the ratio of fuel to air changes slightly, we could have problems. There is no room for errors. What happens if the blower wheel accumulates some dirt, the linkages slip or even if the boiler room temperature changes by 20^0F? If the amount of air supplied to the burner drops, we could soot our boiler, which will be dangerous and increase our fuel costs. (In reality, our burner does not require air but actually requires oxygen. To introduce oxygen into our burner, we need to bring in air since it contains oxygen. *Did you know that for every cubic foot of oxygen that the burner requires, we need to bring in 4.78 cubic feet of air?*

Since we cannot be there 24 hours a day to monitor our burner fuel to air ratio, we decide to add a little extra air as a safety factor. After all, we need to go home or eat at some time. We increase the amount of air from 10 to 13 parts of air. Our new air to fuel ratio is now 13 parts of air to every part of gas. We now have 30% extra or "Excess Air". Our burner efficiency is sacrificed slightly but we are able to have a system that heats without having to be monitored all the time. Even if the blower wheel gets dirty or the linkages slip slightly, we will have a safe boiler. On commercial boilers with power burners, they are designed to supply between 12 and 15 parts of air for each part of gas. The extra air is called "Excess Air" The above burner with 12-15 parts of air would have 20 – 50% excess air. In some industrial facilities, they can reduce the fuel to air ratio even lower than 20% by using an oxygen or O2 trim system. This system uses a sensor in the flue to monitor oxygen content while the burner is running and adjust the air damper.

Atmospheric boilers use a slightly higher amount of air. The primary air for an atmospheric boiler is through the base. The draft diverter is used for dilution air or secondary air. This secondary air is fed into the draft diverter so that it does not steal the heat from the boiler.

A combustion analysis should be performed a minimum of once per year. It does not matter when it is done if it is not done correctly. The picture above shows the hole for the combustion analyzer probe was installed above the draft diverter. This would give you a false reading and could lead to sooting of the boiler.

When is the best time to perform the combustion analysis: Fall, Winter or Spring? To get the proper readings, the boiler should run for 15 minutes prior to any adjustments. This allows the flame to stabilize. When performing the combustion analysis in the Fall and Spring, the boiler may shut off before any adjustments could be completed. The best time to perform the combustion analysis is when the building is under a heavy load, such as mid winter.

A combustion analysis adjusts the optimum amount of air for your boiler and burner. Too much air and you are wasting fuel. Too little and you could soot the boiler. A combustion analysis should also be performed if the following items are done: clean or replace blower wheel or motor, replace gas valve or gas pressure regulator or the linkages slipped.

Fuel oil requires 1,400 cubic feet of air for each gallon of fuel, rated at 140,000 Btuh, for perfect efficiency.

What efficiency should we adjust the burner for?
When starting a boiler, I reference the stated efficiency of the boiler and feel that if I am within one percent of the published efficiency, I will be satisfied with that. When a boiler is tested by an independent organization for its published efficiency, it is under optimum conditions. In reality, the conditions on the job are not the same as the one at the testing lab.

Combustion Air Required For Boilers

Atmospheric with Draft Hood	
Cubic Feet Air per Cubic Foot Gas	Description
10	Perfect Combustion
5	Primary Air (50% excess air)
15	Air for Draft Hood (Secondary Air)
30	Total Air Required

Boiler With Power Burner & Barometric Damper	
Cubic Feet Air per Cubic Foot Gas	Description
10	Perfect Combustion
2	Primary Air (20% excess air)
3	Air for Barometric Damper (Secondary Air)
15	Total Air Required

Boiler With Power Burner & NO Barometric Damper	
Cubic Feet Air per Cubic Foot Gas	Description
10	Perfect Combustion
2	Primary Air (20% excess air)
12	Total Air Required

Combustion Adjustment

We had a new client that was referred to us. The building owner had to rebuild the refractory and replace several boiler tubes each year. When we checked the combustion efficiency, the boiler was reading about **88%** efficient at high fire and almost **90%** efficient at low fire. The boiler was rated for **80%** efficient from the manufacturer. The previous service technician had reduced the firing rate below the condensing point. The acids in the flue gases were destroying the boiler when they condensed. We explained the problem to the owner and suggested that we adjust the efficiency to the proper settings. The owner asked us to keep the settings as they were and he said that he would rather sacrifice some tubes than to increase his fuel costs.

Combustion Analysis Misc. Items

- Do not forget to seal the hole for the combustion analysis probe!
- Mark your setting on linkages. This may help in the future if the linkages slip. A permanent marker is good for this.

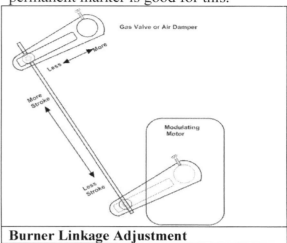

Burner Linkage Adjustment

A Note on Direct Vent Boilers

When you are adjusting the fuel to air ratio on a burner with the combustion air directly vented from the outside, you should be aware of the outside temperature at which you are working. If the weather is very cold when you are adjusting the fuel to air ratio, please be aware that you will have less air when the weather gets warmer due to the air's density.

A note when starting the combustion analyzer:

When starting the analyzer, it will calibrate itself to the conditions found at the probe. If you are checking the combustion efficiency of a non-condensing boiler that uses air from the boiler room for the combustion air, the probe should be calibrated in the same space as the air that will be fed to the boiler and or burner. Do not insert the probe into the stack or in the outside air louver for calibration. The sensing probe of the analyzer will note the temperature at the probe. If the unit is calibrated outside and the boiler uses room air for combustion, your combustion efficiency readings will be wrong.

When testing the fuel to air ratio of a condensing boiler that uses a pipe from the outside for combustion air, insert the probe inside the combustion air pipe as close as possible to the boiler. If your boiler room has both type boilers, you will need to calibrate the analyzer prior to adjusting each boiler.

Type of Analyzer

The older type combustion analyzers used to measure carbon dioxide or CO2. The drawback to measuring only CO2 was that the burner could be producing either carbon monoxide or oxygen and the analyzer would not display that. The analyzer should display oxygen, carbon dioxide and carbon monoxide. In that way you can be assured the burner is firing safely and correctly. If you look at the theoretical Stochiometric ratio of the combustion process, you will see that if you only measure carbon dioxide, you could have carbon monoxide instead of oxygen.

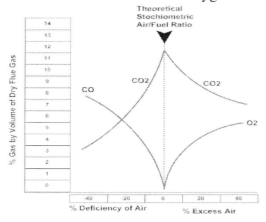

Effect of Air and Barometric Pressure on Burner Combustion.

The following chart displays the effect that the primary air temperature and barometric pressure has on the fuel to air ratio of a burner. In the chart below, we will assume that the burner combustion was originally set up at the following conditions:

Room Temperature 80^0F
Barometric Pressure 29
Burner Excess Air 15%

If any of these conditions change, the fuel to air ratio changes. For instance, if the boiler room temperature drops from 80^0F to 60^0F, the excess air rises from 15% to 20.2%. This is an increase of 34%.

Can you imagine the wide variations that occur in a burner that has the combustion air directly vented from the outside?

Effect of Air Temperature at Same Barometric Pressure		
Combustion Air Temperature	Barometric Pressure	Excess Air %
40	29	25.5
60	29	20.2
80*	**29***	**15***
100	29	9.6
120	29	1.1
Effect of Barometric Changes at Same Temperature		
Combustion Air Temperature	Barometric Pressure	Excess Air %
80	27	7
80	28	11
80*	**29***	**15***
80	30	19
Effect of Temperature and Barometric Changes		
Combustion Air Temperature	Barometric Pressure	Excess Air %
40	31	34.5
60	30	25
80*	**29***	**15***
100	28	5
120	27	-5.5
*Burner Setup Conditions = 80 Degrees ambient air, 15% Excess Air & 29 Barometric Pressure		

Percent of Air That will Permit Combustion

	Minimum % of Air	Maximum % of Air
Natural Gas	64	247
Oil	30	173
Coal Pulverized	8	425

Combustion Analysis Report

Use an analyzer that has a print out to verify your findings. I would also suggest that you make a copy of your report. The thermal printouts from the analyzers have a tendency to fade over the years. This should be filed into the customer file. The analyzer should be calibrated yearly. The report should include the following:

Boiler & Burner Manufacturer
Boiler & Burner Model
Serial Number
Age
Input
Output
Temperature Ambient and Flue Gas
CO PPM (Air Free)
Excess Air Percentage
O2 Percentage
CO2 Percentage
Efficiency

Always keep the startup report in the client's permanent records to protect yourself.

Ignition Temperature of Fuels

Fuel	Degrees F
Kerosene	500
Light Fuel Oil	600
Gasoline	735
Butane-N	760
Heavy Fuel Oil	765
Coal	850
Propane	875
Methane	1076
Hydrogen	1095
Natural Gas	1163
Carbon Monoxide	1170
Natural Gas	1200

Typical Combustion Test Results
Category 1 Boilers
Atmospheric Gas Burner

Oxygen	7% - 9%
Stack Temperature	325 to 500 Degrees F
Draft in Water Column Inches	-.0 " WC to -.04" WC
Carbon Monoxide PPM	<100 PPM Air Free

Gas Power Burner

Oxygen	3% - 6%
Stack Temperature	275 to 500 Degrees F
Draft in Water Column Inches	-.02" WC to -.04" WC
Carbon Monoxide PPM	<100 PPM Air Free

Typical Combustion Test Results
Category 1 Boilers
Oil Power Burner

Oxygen	4% - 7%
Stack Temperature	325 to 600 Degrees F
Draft in Water Column Inches	-.04" WC to -.06" WC
Carbon Monoxide PPM	<100 PPM Air Free with 0 Smoke

CO Air Free

When performing a boiler combustion analysis, the measurement of carbon monoxide or CO is an important factor to check and adjust. If carbon monoxide is present in the flue gases, it indicates incomplete combustion. The industry guideline for carbon monoxide in flue gases is 400-ppm "air free". What is CO "Air-Free"?

Let us look at a hypothetical problem. We are performing a service call on a boiler and we insert a probe into the stack to measure the carbon monoxide in the stack. Our initial reading is 350-ppm carbon monoxide and the burner has 7% oxygen. This reading is called "as-measured". We feel good that it is below the 400 ppm that is allowed in the flue. When we look at the CO air free reading, we actually have

526 ppm CO. We still have some adjustments to perform to get the CO levels down. Adding more air will not solve the problem. It will simply mask it. To get the true reading of carbon monoxide we have the following formula:

$$CO\ Air\ Free = 20.9 / (20.9 - O2) \times CO$$

CO Air free takes the excess air from the combustion to see the true carbon monoxide reading. Many of the newer analyzers will calculate CO air free for you. To use the chart below, find the as-measured carbon monoxide readings in bold along the top of the chart. Look at the left side to see your measured oxygen in the flue gases. Where the two meet is the actual CO air free readings. For example, 300 ppm at 6% O2 shows 421 ppm CO Air Free.

Diluted or "As Measured" CO Reading							
O2%	200	225	250	275	300	325	350
1.0	210	236	263	289	315	341	368
2.0	221	249	276	304	332	359	387
3.0	234	263	292	321	350	379	409
4.0	247	278	309	340	371	402	433
5.0	263	296	329	361	394	427	460
6.0	281	316	351	386	421	456	491
7.0	301	338	376	413	451	489	526
8.0	324	365	405	446	486	527	567
9.0	351	395	439	483	527	571	615
10.0	383	431	479	527	575	623	671
11.0	422	475	528	581	633	686	739
12	470	528	587	646	704	763	822
13	529	595	661	728	794	860	926
14	606	682	757	833	909	984	1060
15	708	797	886	974	1063	1151	1240

Calculate CO Air Free from CO Readings							
Diluted or "As Measured" CO Reading							
O2%	25	50	75	100	125	150	175
1.0	26	53	79	105	131	158	184
2.0	28	55	83	111	138	166	194
3.0	29	58	88	117	146	175	204
4.0	31	62	93	124	155	186	216
5.0	33	66	99	131	164	197	230
6.0	35	70	105	140	175	210	245
7.0	38	75	113	150	188	226	263
8.0	41	81	122	162	203	243	284
9.0	44	88	132	176	220	263	307
10.0	48	96	144	192	240	288	336
11.0	52	106	158	211	264	317	369
12	59	117	176	235	394	352	411
13	66	132	198	265	331	397	463
14	76	151	227	303	379	454	530
15	89	177	266	354	443	531	620

Boiler Emissions
Due to worries about global climate change, boiler emissions are a growing concern. Many states are requiring reduced emissions from the boilers. Please check with your state to determine their emission limits. In some states, different areas or cities may have different requirements. There are several flue gas by-products that are being regulated.

Sulfur Dioxide
Sulfur Dioxide is also a by-product of combustion. It is rare with gaseous fuels. It is also a primary contributor to acid rain, which causes acidification of streams and lakes. It is released primarily from burning fuels that contain sulfur (such as coal, oil, and diesel fuel). Sulfur dioxide, when combined with water, forms sulfuric acid.

Particulate Matter
Particulate matter, or otherwise known as soot, is mostly formed from incomplete combustion of fuel. It is comprised of unburned fuel, organic chemicals, soil, dust, sulfates, nitrates, oxides and or carbons. There are two basic types of particulate matter, Pm and PM_{10}. PM_{10} is particulate matter that is less that 10 microns in diameter. To see how small this is, consider that a human hair is about 70 microns. The PM_{10} is small enough to bypass the human body's natural

filtering system and imbed themselves in the lungs. It can trigger asthma attacks, coughing and acute bronchitis. It is more prevalent in oil than gas. There are some areas that are trying to test for particulates as small as 2.5 microns

Carbon Dioxide

Carbon dioxide, when combined with water, forms carbonic acid.

Nitrogen Compounds or NO_x

NO_x is principally made up of two components, Nitric Oxide(NO) and Nitrogen Dioxide(NO_2). NO_2, when combined with other pollutants, such as Volatile Organic Compounds (VOCs), is believed to form O3 or ground level ozone and acid rain. NO_x emissions become more prevalent when the flame temperature is above $2,800^0F$ and the fuel to air combustion ratios are between 5-7% O_2. This is typically where the commercial burner manufacturers direct the fuel to air ratios to be set.

There are two types of NOx, Thermal NOx and Fuel NOx. Thermal NOx is the most common. It is produced in the boiler when oxygen and nitrogen combine under elevated temperatures. Fuel NOx rarely occurs when firing with gas. It is common in heavier fuel oils.

Acid Rain and Stack Temperature

The flue gases from a fossil-fueled boiler contain the following: oxygen, carbon dioxide, carbon monoxide, sulfur dioxide and free water. If allowed to condense, these acids will destroy the stack, chimney and boiler. The following are the acid dew point temperatures of the most common fuels. Flue gas condensation could also occur if the burner is under fired. This means that the supply of fuel to the burner is less than the manufacturer's recommendations.

Draft

Draft should be measured and logged during the start up. Excess draft can steal the heat from the boiler and direct it up the stack. This will waste energy. It could also distort the flame pattern, causing Carbon Monoxide CO to form.

Rule of thumb. For every 0.01" w.c. the excess draft can be reduced, the fuel consumption is reduced by 1%

Type of Heating System	Overfire Draft	Stack Draft
Gas, Atmospheric	Not Applicable	-.02 to -.04" WC
Gas, Power Burner	-.02" WC	-.02 to -.04" WC
Oil, Conventional	-.02" WC	-.04 to -.06" WC
Oil, Flame Retention	-.02" WC	-.04 to -.06" WC
Positive Overfire Oil & Gas	+.4 to +.6	-.02 to -.04" WC

Fuel	Acid Dew Point Temperature	Minimum Allowable Stack Temperature
Natural Gas	150	250
#2 Fuel Oil	180	275

Pollution Conversions

After you finish the combustion analysis on your new boiler, you may be asked to forward an emissions report. To calculate that, take the readings and follow the following formulas. To convert PPM to any of the units below, multiply PPM by the number in the correct column and row.

Definitions

Lb/MBTU = pounds of pollutants per million BTU

Mg/NM3 = Milligrams of pollutants per Million BTU

MG/KG = Milligrams of pollutants per Kilogram of fuel burned

G/GJ = Grams of pollutant per Gigajoule

Based on 3% excess Oxygen and dry gas

Fuel	Multiply PPM by factor below				
	Pollutant	Lb/MBTU	MG/NM3	MG/KG	G/GJ
Nat Gas	CO	0.00078	1.249	12.647	0.338
Nat Gas	NOx	0.00129	2.053	20.788	0.556
Nat Gas	SO2	0.00179	2.857	28.949	0.775
Oil #2, #6	CO	0.00081	1.249	15.118	0.354
Oil #2, #6	NOx	0.00134	2.053	24.850	0.582
Oil #2, #6	SO2	0.00186	2.857	34.605	0.811

Boiler Gas Consumption

Btuh	Cu Feet/Hr	Cubic Feet/ Min	Cubic Feet/ Sec
20,000,000	20,000	333.33	5.56
19,000,000	19,000	316.67	5.28
18,000,000	18,000	300.00	5.00
17,000,000	17,000	283.33	4.72
16,000,000	16,000	266.67	4.44
15,000,000	15,000	250.00	4.17
14,000,000	14,000	233.33	3.89
13,000,000	13,000	216.67	3.61
12,000,000	12,000	200.00	3.33
11,000,000	11,000	183.33	3.06
10,000,000	10,000	166.67	2.78
9,000,000	9,000	150.00	2.50
8,000,000	8,000	133.33	2.22
7,000,000	7,000	116.67	1.94
6,000,000	6,000	100.00	1.67
5,000,000	5,000	83.33	1.39
4,000,000	4,000	66.67	1.11
3,000,000	3,000	50.00	0.83
2,000,000	2,000	33.33	0.56
1,000,000	1,000	16.67	0.28
Based upon 1,000 Btu's per cubic foot			

Did You Know…
A cubic inch of water evaporated under ordinary atmospheric pressure (14.7 psig) will be converted into approximately one cubic foot of steam.

Did You Know…
That the computing power in a greeting card that sings "Happy Birthday" has more computing power than existed in the entire world before 1950?

Chapter 20
Clocking a Gas Meter

Many engineers specify that the new boilers be "clocked" to assure proper firing rate. Clocking a gas meter is a way of verifying the actual firing rate of the boiler. There are a couple of ways to perform that task. To properly "clock" a meter, be sure that your boiler is the only apparatus that is firing. The boiler should also be at high fire. A way to check this is to shut off all the items in the boiler room and observe the gas meter. If the hands on the dial are still moving, there are other items consuming gas. It is sometimes difficult to perform this task in a commercial building, as there are could be many items operating simultaneously.

Some commercial gas meters require you to compensate for the temperature as well as the pressure of the gas. When the meter was calibrated, it was at a certain temperature and gas pressure at the factory. The field conditions may be different. In the end of the chapter, there are some figures that will help you to compensate for the conditions found at the job site.

How to Clock the Meter
Once you are sure that there are no other appliances using gas, you will need to start your unit and assure that it is firing at full rate. You then want to count the number of revolutions the most sensitive dial on the gas meter makes in one minute. Most natural gas has a heating value of 1,000 BTU/cubic foot. Let us assume that our most sensitive dial is ½ cubic feet per revolution.

A. Count the revolutions the ½ cubic foot dial makes in one minute.
B. Multiply the revolutions by 30,000 to obtain the firing rate in Btu's/ Hr

For example, the ½ cubic foot dial made 3.2 revolutions in one minute. The boiler is firing at 3.2 revolutions x 30,000 BTU/revolution = 96,000 BTUH. If you find that the heating value is different from 1,000 Btu's per

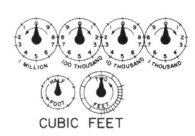

cubic foot, you would have to make an adjustment. The local gas company could inform you of the heating content of their gas. For example, if the gas company tells you that the heating value is 1,050 BTU/ cubic foot, you would need to adjust your final reading. 1,050 BTU/ Cubic foot (Actual BTU) divided by 1,000 Btu/ Cubic foot (This was assumed to be the BTU content) = 1.05. Therefore, to recalculate the new rate, we would multiply 96,000 Btuh (From above) x 1.050 = 100,800 Btu/ HR. This is the actual firing rate of the appliance.

The 30,000 calculation only works with ½ cubic foot dial. For other size dials, see below.
Remember our basic formula is Number of revolutions x factor below = BTU/ Hr. This is based on 1,000 BTU/ Cubic Foot.
NOTE: To get a more accurate reading, it is better to allow the test to be done for a longer time. I would recommend 5 minutes. You would then divide the reading by 5 to get the average. The chart below features different timing for the dials, up to five minutes. For example, if the 5 "Cubic Feet per Revolution" dial made two revolutions in five minutes, your firing rate would be as follows:
60,000 x 2 = 120,000 Btuh

Multiplying Factor for Gas Meter

Cubic Feet per Revolution	1 Minute Timing	2 Minute Timing	3 Minute Timing	5 Minute Timing
	BTUH			
½	30,000	15,000	7,500	6,000
1	60,000	30,000	15,000	12,000
2	120,000	60,000	30,000	24,000
5	300,000	150,000	75,000	60,000
Based on 1,000 Btu per cubic foot of gas				

Clocking a Gas Meter Option 2

A second method for "clocking" a gas meter is as follows

Start the boiler; making certain that no other gas-fired appliance is operating. Measure the amount of time it takes for the smallest dial to make one complete revolution. In the above dials, the ½ cubic foot dial is the timing dial.

Refer to a natural gas timing chart under ½ cubic foot column and see what the input is to your boiler.

Check and compare the calculated input with the input rating on the heating unit data plate. If the unit is under-fired or over-fired by more than 10%, check the gas pressure to the unit with a fluid filled manometer and adjust as necessary.

(For example, the unit being tested takes 29 seconds for the ½ cubic foot dial to make one complete revolution. Using the chart, this translates to 62 cubic feet per hour. Based upon the assumption that one cubic foot of natural gas has 1,000 BTU's (Check with your local utility for actual BTU content), the calculated input is 62,000 BTU's per hour.

You will get a better reading by allowing the dial to rotate several times and dividing the total by the amount of revolutions to get an average. In the above example, if it took 1 minute, 27 seconds or 87 seconds to make three revolutions, our average input would be 29 seconds.

Natural Gas Timing Chart in Cubic Feet/Hour

Seconds for one revolution	1/2 Cu Ft	1 Cu Ft	2 Cu Ft	5 Cu Ft
10	180	360	720	1,800
11	164	327	655	1,636
12	150	300	600	1,500
13	138	277	555	1,385
14	129	257	514	1,285
15	120	240	480	1,200
16	112	225	450	1,125
17	106	212	424	1,059
18	100	200	400	1,000
19	95	189	379	947
20	90	180	360	900
21	86	171	345	857
22	82	164	327	818
23	78	157	313	783
24	75	150	300	750
25	72	144	288	720
26	69	138	277	692
27	67	133	267	667
28	64	129	257	643
29	62	124	248	621
30	60	120	240	600
31	58	116	232	581
32	56	113	225	563
33	55	109	218	545
34	53	106	212	529
35	51	103	205	514
36	50	100	200	500
37	49	97	195	486
38	47	95	189	474
39	46	92	185	462
40	45	90	180	450
40	45	90	180	450

Seconds for one revolution	1/2 Cu Ft	One Cu Ft	Two Cu Ft	Five Cu Ft
41	44	88	176	440
42	43	86	172	430
43	42	84	167	420
44	41	82	164	410
45	40	80	160	400
46	39	78	157	391
47	38	77	153	383
48	37	75	150	375
49	37	73	147	367
50	36	72	144	360
51	35	71	141	353
52	35	69	138	346
53	34	68	136	340
54	33	67	133	333
55	33	65	131	327
56	32	64	129	321
57	32	63	126	316
58	31	62	124	310
59	30	61	122	305
60	30	60	120	300
62	29	58	116	290
64	29	56	112	281
66	29	54	109	273
68	28	53	106	265
70	26	51	103	257
72	25	50	100	250
74	24	48	97	243
76	24	47	95	237
78	23	46	92	231
80	22	45	90	225
82	22	44	88	220
84	21	43	86	214
86	21	42	84	209
88	20	41	82	205

NOTES:

On a commercial gas meter, you may have to calculate a pressure and/or temperature correction factor. You will need to contact the local gas company for this factor.

Gas Pressure Correction Factor

Actual Meter Pressure (psi)	Meter Base Pressure				
	4 oz or 7" w.c.	8oz or 14" w.c.	10 oz or 17.5" w.c.	1 Psi or 28" w.c.	2 psi or 56" w.c.
0	0.983	0.966	0.958	0.935	0.878
1/4	1.0	0.983	0.975	0.951	0.893
1/2	1.017	1	0.992	0.968	0.909
5/8	1.026	1.008	1	0.976	0.916
1	1.051	1.034	1.025	1.000	0.929
2	1.119	1.101	1.092	1.065	1.000
3	1.188	1.168	1.158	1.130	1.061
4	1.256	1.235	1.225	1.195	1.122
5	1.324	1.302	1.291	1.260	1.183
6	1.392	1.369	1.358	1.325	1.244
7	1.461	1.436	1.424	1.390	1.305
8	1.529	1.503	1.491	1.455	1.366
9	1.597	1.570	1.557	1.520	1.427
10	1.666	1.638	1.624	1.584	1.488

Gas Temperature Correction Factor

Gas Temperature Degrees F	Meter Calibration Temperature				
	60 Degrees F	65 Degrees F	68 Degrees F	70 Degrees F	72 Degrees F
0	1.130	1.141	1.148	1.152	1.157
5	1.118	1.129	1.135	1.140	1.144
10	1.106	1.117	1.123	1.128	1.132
15	1.095	1.105	1.112	1.116	1.120
20	1.083	1.094	1.100	1.104	1.108
25	1.072	1.082	1.089	1.093	1.097
30	1.061	1.071	1.078	1.082	1.086
35	1.051	1.061	1.067	1.071	1.075
40	1.040	1.050	1.056	1.060	1.064
45	1.030	1.040	1.046	1.050	1.053
50	1.020	1.029	1.035	1.039	1.043
55	1.010	1.019	1.025	1.029	1.033
60	1.000	1.010	1.015	1.019	1.023
65	0.990	1.000	1.006	1.010	1.013
70	0.981	0.991	0.996	1.000	1.004
75	0.972	0.981	0.987	0.991	0.994
80	0.963	0.972	0.978	0.981	0.985
85	0.954	0.963	0.969	0.972	0.976
90	0.945	0.955	0.960	0.964	0.967
95	0.937	0.946	0.951	0.955	0.959
100	0.929	0.938	0.943	0.946	0.950

Chapter 21
Differentiating Your Company

What happens if there is a problem on the job?

In spite of all your due diligence, planning and expert installation, there is a problem on the job. How do you handle it? I know a very successful contractor that once told me, "I have never had a problem job." I looked at him incredulously and asked how that was. He told me that he does whatever it takes to make the job right. Some jobs just cost him more than others did. He said that he learned long ago that it does not pay to walk away from the problem job.

Thermometers

One of the items you seldom see on a boiler replacement project now is a stack thermometer. It used to be quite common to see one in the stack. It makes a more professional job and it is a great place to insert your combustion analyzer probe. It is also useful in diagnosing system problems. If the stack temperature starts to rise, you will know that something is wrong and should be addressed.

Under Promise, Over Deliver

I have always been a fan of this adage. Consider the following scenario. In the first, you tell the client that they will save 25% on their heating costs by having your organization install the boiler. When you calculate the savings, you find out that the client saved 20%. In the next scenario, you tell the client that you think they will save about 10% on their heating costs. When you calculate the savings, it comes to a 20% savings. In the first one, your client is angry and wonders if they can trust you. In the second one, you are a hero. You saved twice as much as you said. The savings were equal in both. Always under promise and over deliver.

Painting the piping

If you want to really make your project stand out, consider painting the gas piping.

Pipe Labels

The sign of a true professional project is when the installer applies descriptive labels on the piping along with arrows. It certainly helps future service techs to diagnose a problem. One note, please make sure the arrows are going the right way. *Sorry that this has to be said.*

Leave your sticker on the equipment

The client may soon forget who installed their new boiler. Always place a sticker with your telephone number on the equipment

Service Agreement

Right after the job is complete, is when the customer should be most satisfied. It is then that you should present a service agreement proposal to the owner. Most owners do not realize that preventive maintenance is not included under the boiler warranty.

Did Client Actually Save Money?

If you spent as much money as your client did on the new heating system, would you not want to know if you did indeed save money? You could ask for copies of the last two years fuel costs and perform a quick survey. This will make you stand out to the client. You can also use it as a sales tool. For instance, if the client's fuel costs dropped 20%, you have a legitimate testimonial for future sales. You can usually download the degree-days for your area to help you with this study.

Maintenance Manuals

Do not forget to make copies of the maintenance manuals for the customer's records.

Clean Up After Yourself

One thing that really annoys most clients is when the technician leaves a mess. Encourage your employees to clean up after themselves. Remember to clean the new boiler as well. They sometimes will get solder splatter or dirt on them after the installation. The client is thinking of how much money they spent and a clean boiler looks better. Try to imagine what you would feel like if you purchased a new auto and found smudged handprints and scratches on it.

Ask for a letter of recommendation
As I stated earlier, when the job is complete is when the customer is happiest. You could ask for a letter of recommendation so that you could show it to potential customers.

Send a Thank You Note
It is something that will separate you from the other companies. I have had people tell me that they have kept handwritten thank you notes for years. Many installers look at an installation as a one-time relationship. I think it could be like an annuity. If you treat the client right, they will use your company for service year after year. In addition, your client will know others in the same position at other organizations. They may recommend you to their colleagues.

Stop back next winter
One way to make your organization stand out from the competition is to stop back and see the client the following winter. Most companies cannot wait to get out of the job and forget about the prior customers. It will reassure the client that choosing your company was the right decision. Remember, your client probably knows others in the same position at different facilities.

Add Lights in Boiler Room
One of the ways to really have your company stand out is to add lights by your new boilers. A single shop light is very reasonable. It will certainly separate your firm. The future service technician will thank you as well.

Misc Things…
The cost to operate a boiler is usually 4 times the installation cost.
According to Consortium for Energy Efficiency, condensing boilers make up 20-25% of the commercial boiler market.
Did you know that for every 40 degree F increase in combustion air, the boiler efficiency increases 1%?

Did you know that for every 40 degree F decrease in flue gas temperature, the boiler efficiency increases 1%?

What happens when boiler room floods
When a boiler room floods, all components that were under water should be replaced. Consult with the manufacturer as to their requirements when a flood occurs. If the building is prone to flooding, you may want to consider a higher pad. Some sensors can be mounted on the floor that will detect water and could alarm.

A Note on Valves
Gate and globe valves will freeze open in some instances. When opening a gate or globe valve, always open the valve all the way and then close the valve 1/2 turn. This will reduce the chances of the valve stuck open. In some instances, the gate valve could freeze closed. When opening the valve, the gate could be stuck inside the valve. This is a common problem with old gate valves.

Types of Heat Transfer
Conduction is the transfer of heat through a material or substance. It could even transfer to the adjoining material. An example of this is the heat that transfers from the frying pan on a stove to the handle of the pan. In our industry, it is like the heat that transfers along a pipe as it is soldered. Heat is also conducted through the ceilings, walls and floors of homes.

Convection is the transfer of heat by a liquid or gas (such as air). Circulatory air motion due to warmer air rising and cooler air falling is a common mechanism by which thermal energy is transferred. In our industry, it would be the heat that is circulated in a room heated with fin tube radiation. Convective heat loss also occurs through cracks and holes in the building and gaps and voids in ceilings, walls, and floors—and in the insulation.

Radiation heat transfer occurs between objects that are not touching. The most common example of this is the way the sun heats the earth. The sun warms the earth without warming the space between the sun and the earth. An example in our industry is the heat that you feel from a radiant heater or large cast iron radiator.

Types of Heat Transfer	
Sensible Heat	Sensible heat is any heat transfer that causes a change in temperature without causing a change of state. Sensible heat can be measured with a dry bulb thermometer. For example, when the temperature is increased over a heating coil, the temperature differential is sensible heat.
Latent Heat	Latent heat is the amount of heat required to cause a change of state. In a boiler system, this would be the amount of heat added to water to cause it to change from water to steam. It requires 970.4 Btus to raise 1 pound of water at 212 degrees f to 1pound of steam. The latent heat that is added to change water to steam is also given back when the steam condenses in the radiator or coil.
Total Heat	Total heat is the sum of the sensible and latent heat in an exchange process. It is sometimes called enthalpy.

Did You Know...
As of 2009, between 7 and 7.8 million replacement water heaters are sold every year?
US Dept of Energy

Did You Know...
That the half-life of carbon monoxide poisoning is 5 hours?
It take 5 hours for CO levels in the body to drop to half its concentration once the exposure has ended.

Chapter 22
Heating Formulas & Rules of Thumb

Combustion Air Openings

Each fuel-burning piece of equipment requires combustion air to operate safely. The following are some guidelines to help you see whether the existing combustion air louvers will be adequate for the replacement project.

Number of openings required = 2

- Each boiler room should have two openings. One should be within one foot of ceiling and the other opening within one foot of the floor.

Size of Direct Openings

- 1" Free area for each 4,000 Btuh

Horizontal Openings

- 1" Free space for each 2,000 Btuh

Vertical Openings

- 1" Free space for each 4,000 Btuh

Mechanical Ventilation

1 cfm per 2,400 Btuh

Steam Systems

Steam Pipe Sizes Steam Main

Pipe Size	Btu/Hr	Pipe Size	Btu/Hr
2"	155,520	4"	912,000
2 ½"	247,680	5"	1,612,800
3"	446,400	6"	2,707,200
3 ½"	643,200	8"	5,347,200

Sizing a Dry Return Piping System

Pipe Size	Btu/Hr	Pipe Size	Btu/Hr
1"	98,800	2 ½"	1,180,800
1 ¼"	208,320	3"	2,160,000
1 ½"	326,400	4"	4,636,800
2"	710,400		

Size Steam Main by Connected Radiation

Radiation Sq Feet	Pipe Size Inches
75-125	1 ¼
125-175	1 ½
175-300	2
300-475	2 ½
475 – 700	3
700-1200	4
1200-1975	5
1975 – 2850	6

Boiler Feed Sizing Rules of Thumbs

Boiler Feed Sizing Rules of Thumbs	
Pump GPM =	Evaporation Rate x 2
Tank Sizing =	Pump GPM x 20*
*Based on 20 minutes storage. One gallon storage per Boiler HP	

Condensate Pump Sizing Rules of Thumb

Condensate Pump Sizing Rules of Thumb	
Pump GPM =	Evaporation Rate x 3
Tank Sizing =	Pump GPM x 1

Steam and Condensate Rules of Thumb

Lbs. Steam / Hr =	Boiler Hp x 34.5
	Btuh divided by 960
Evaporation Rate in GPM =	Boiler Hp x .069
	Lbs. Steam Hr /500
EDR =	Boiler Hp x 139
	Btu/Hr /240 Steam
	Btu/Hr / 150 Water
	0.000496 GPM
	0.25 Pounds Steam
Evaporation Rate in GPM	EDR/1000 x 0.5
	Lbs/Steam/ Hr /500
Lb Steam condensate/ hr =	EDR divided by 4
Boiler Evap. Rate =	Boiler HP x .069

All boilers have an evaporation rate of 1/2 gpm per 240,000 Btuh boiler rating. Feedwater should be fed at a rate of 1 gpm or more per 240,000 Btuh

A cubic inch of water evaporated under ordinary atmospheric pressure (14.7 psig) will be converted into approximately one cubic foot of steam.

One cubic foot of steam exerts a mechanical force equal to that needed to lift 1,955 pounds one foot.

Each nominal boiler horsepower requires 4-6 gallons of water per hour.

Boiler Pressure Designations

Low Pressure	Steam up to 15 psig Hot Water Up to 160 psig & 250° F
Medium Pressure	Steam pressure up to 50 psig
High Pressure	Steam pressure greater than 50 psig *Some locations categorize anything over 15 pounds as high pressure.

Boiler Ratings

Boiler HP	Btu/hr Output	Steam Lb/hr	EDR Rating Sq Feet
20	670,000	690	2,790
30	1,005,000	1,035	4,185
40	1,340,000	1,380	5,580
50	1,675,000	1,725	6,975
60	2,010,000	2,070	8,370
70	2,345,000	2,415	9,765
80	2,680,000	2,760	11,160
100	3,350,000	3,450	13,950
125	4,185,000	4,313	14,438
150	5,025,000	5,175	20,935
200	6,695,000	6,900	27,915
250	8,370,000	8,625	34,895
300	10,045,000	10,350	41,875
350	11,720,000	12,075	48,825
400	13,390,000	13,800	55,920
450	15,064,000	15,520	63,000
500	16,740,000	17,250	69,790
600	20,085,000	20,700	83,750
650	21,759,000	22,425	91,000
700	23,432,000	24,150	98,000
750	25,106,000	25,875	105,000
800	26,780,000	27,600	112,000
1000	33,475,000	34,500	140,000

Boiling Temperatures of Water at Different Altitudes F

Altitude Feet	Gage Pressure				
	0	1	5	10	15
-500	212.8	216.1	227.7	239.9	250.2
-100	212.3	215.5	227.2	239.4	249.9
Sea Level	212.0	215.3	227.0	239.3	249.7
500	211.0	214.4	226.3	238.7	249.2
1,000	210.1	213.5	225.5	238.1	248.6
1,500	209.4	212.7	225.0	237.6	248.2
2,000	208.2	211.7	224.1	236.8	247.7
2,500	207.3	210.9	223.4	236.3	247.2
3,000	206.4	210.1	222.7	235.7	246.7
3,500	205.5	209.2	222.1	235.1	246.2
4,000	204.7	208.4	221.4	234.6	245.7
4,500	203.7	207.5	220.7	234.0	245.2
5,000	202.7	206.8	218.7	233.4	244.7
6,000	200.9	205.0	217.3	232.4	243.8
7,000	199.1	203.3	216.1	231.3	242.9
8,000	197.4	201.6	214.8	230.3	242.0
9,000	195.7	200.0	213.5	229.3	241.3
10,000	194.0	198.4	212.3	228.3	240.4

	Boiling Temperatures at Various Pressures			
Vacuum In Hg.	**Boiling Temp.**		**Vacuum In Hg.**	**Boiling Temp.**
29	76.62		6	200.96
28	99.93		5	202.25
27	114.22		4	204.85
26	124.77		3	206.70
25	133.22		2	208.5
24	140.31		1	210.25
23	146.45		**Gage Pounds**	**Boiling Temp.**
22	151.87		0	212
21	156.75		1	215.6
20	161.19		2	218.5
19	165.24		3	221.45
18	169.0		4	224.4
17	172.51		5	227.10
16	175.8		6	229.8
15	178.91		7	232.3
14	181.82		8	234.8
13	184.61		9	237.3
12	187.21		10	239.4
11	189.75		11	241.5
10	192.19		12	243.6
9	194.5		13	245.7
8	196.73		14	247.8
7	198.87		15	249.8

Properties of Saturated Steam			
Absolute Pressure	Gage Pressure	Temp F	Volume of 1lb steam cu.ft.
14.7	0	212.0	26.8
15.7	1	215.5	25.2
16.7	2	218.7	23.5
17.7	3	221.7	22.3
18.7	4	224.5	21.4
19.7	5	227.3	20.1
20.7	6	229.9	19.4
21.7	7	232.4	18.7
22.7	8	234.9	18.4
23.7	9	237.2	17.1
24.7	10	239.5	16.5
25.7	11	241.7	16
26.7	12	243.8	15.3
27.7	13	245.9	15
28.7	14	247.9	14.3
29.7	15	249.8	14

Properties of Saturated Steam			
Gage Pressure	Sensible Heat	Latent Heat	Total Heat
0	180	970	1150
1	183	968	1151
2	187	966	1153
3	190	964	1154
4	192	962	1154
5	195	960	1155
6	198	959	1157
7	200	957	1157
8	201	956	1157
9	205	954	1159
10	207	953	1160
11	209	951	1160
12	212	949	1161
13	214	948	1162
14	216	947	1163
15	218	945	1163

Steam Capacity per Boiler HP

Feed Water Temperature	Pounds of dry saturated steam per boiler HP @ system pressure (PSIG) at a given feed water temperature (F0)			
	Steam Pressure			
Feed Water Temperature	0	2	10	15
100	30.9	30.8	30.6	30.6
110	31.2	31.2	30.9	30.8
120	31.5	31.4	31.2	31.1
130	31.8	31.7	31.5	31.4
140	32.1	32	31.8	31.7
150	32.4	32.4	32.1	32.0
160	32.7	32.7	32.4	32.4
170	33.0	33.0	32.7	32.6
180	33.4	33.3	33.0	32.9
190	33.8	33.7	33.4	33.3
200	34.1	34.0	33.7	33.6
212	34.5	34.4	34.2	34.1
220	34.8	34.7	34.4	34.3
227	35.0	34.9	34.7	34.5
230	35.2	35.0	34.8	34.7

Typical Water Quality Parameters in Steam Systems

Feedwater	Softness	Less than 1 ppm
Boiler Water	Hardness	Less than 1 ppm
Boiler Water	Ph	7-9 or 9.5-11*
Boiler Water	TDS	1,500-3000 ppm 2,000-4000 microsiemens
Boiler Water	Sulfite	30-60 ppm
Boiler Water	Hydroxyl Alkalinity	200 - 400 ppm
Condensate	pH	8.2-9.0
Condensate	TDS	Less than 20 ppm
*Verify with boiler manufacturer		

Water Treatment
pH Scale

10,000,000	PH = 0	Battery Acid
1,000,000	PH = 1	Hydrochloric Acid
100,000	PH = 2	Lemon Juice, Vinegar
10,000	PH = 3	Grapefruit, Orange Juice
1,000	PH = 4	Acid Rain, Tomato Juice
100	PH = 5	Black Coffee
10	PH = 6	Urine, Saliva
1	PH = 7	"Pure" Water
1/10	PH = 8	Sea Water
1/100	PH = 9	Baking Soda
1/1,000	PH = 10	Milk of Magnesia
1/10,000	PH = 11	Ammonia Solution
1/100,000	PH = 12	Soapy Water
1/1,000,000	PH = 13	Bleaches, Oven Cleaner
1/10,000,000	PH = 14	Liquid Drain Cleaner

Hydronic Information
Pipe Flow Rates

Copper Pipe Maximum Hydronic Flow Rates
Based on 20 degree F Delta T

Pipe Size	Maximum Flow GPM	Btuh
½"	1 1/2	15,000
¾"	4	40,000
1"	8	80,000
1 ¼"	14	140,000
1 ½"	22	220,000
2"	45	450,000
2 ½"	85	850,000
3"	130	1,300,000

Steel Pipe Maximum Hydronic Flow Rates
Based on 20 degree F Delta T

Pipe Size	Maximum Flow GPM	Btuh
½"	2	15,000
¾"	4	40,000
1"	8	80,000
1 ¼"	16	140,000
1 ½"	25	220,000
2"	50	450,000
2 ½"	80	850,000
3"	140	1,300,000
4"	300	3,000,000
5"	550	5,500,000
6"	850	8,500,000
8"	1,800	18,000,000
10"	3,200	32,000,000
12"	5,000	50,000,000

PEX Piping Maximum Hydronic Flow Rates

Pipe Size	3/8"	½"	5/8"	¾"	1"
Max GPM	1.2	2	4	6	9.5
BTUH	12,000	20,000	40,000	60,000	95,000

Hydronic Boilers Rules of Thumb
These are rules of thumb. Please check with the boiler manufacturer for actual requirements.

Typical Hydronic Velocity = 1.5 -4.5 Feet per second in occupied areas. Slightly higher in unoccupied areas. Flows above 6 feet per second could erode copper

To find Fluid Velocity

$$Feet\ per\ Second\ \frac{0.408\ x\ GPM}{(Pipe\ Diameter\ Inches)^2}$$

One gpm will deliver approximately 10,000 Btuh with a 20 degree F delta T

Maximum flow through a boiler is boiler output divided by the temperature rise* divided by 500. *Please note most boilers require a 20-30 degree temperature rise.*

3.45 Gpm/BHP @ 20-degree delta T
2.30 GPM/BHP @ 30-degree delta T
10,000 Btuh/GPM @ 20-degree Delta t
15,000 Btuh /GPM @30-degree Delta T

Hydronic Heat Flow Equation
Q=500 x GPM x Δt
Q = Btuh
GPM = Gallons per minute
Δt = Temperature difference F^0

Hot Water System Makeup:
Minimum connection size shall be 10% of largest system pipe or 1", whichever is greater 20" pipe should equal a 2" connection.

Sizing a Circulator
Flow rate

$$GPM = \frac{BTUH}{\triangle °F\ x\ 500}$$

$$GPM = \frac{BTUH}{8.33\ x\ 60\ x\ \triangle °F}$$

8.33 (weight of 1# water)
60 (Minutes)
△°F Temperature rise (Usually 20 degrees F)

Most round up to 10,000 Btu/ gpm
For example, 900,000 Btuh output divided by 10,000
= 90 gpm

Calculate pump head
1 Measure longest run in feet
2 Multiply by 1.5 to calculate fittings and valves
3 Multiply by 0.04 (4' head for each 100' of pipe
ensures quiet operation)
For example, 100 feet is longest run
100 x 1.5 x .04 = 6 feet of head

WATER FORMULAS
Water Density ^{v}f

Temperature	Density		Specific Volume
^{0}F	Lb/ft^3	$Lb/gallon$	^{v}f
32	63.41	8.48	0.01747
39	63.42	8.48	0.01602
50	63.40	8.48	0.01602
68	63.31	8.46	0.01605
86	63.15	8.44	0.01609
122	62.67	8.38	0.01621
140	62.35	8.34	0.01629
158	62.01	8.29	0.01638
176	61.63	8.24	0.01648
194	61.22	8.18	0.01659
212	60.78	8.13	0.01671

Typical Water Density	
One Cubic Foot Water =	62.43 lbs
	7.48 gallons
	29.92 quarts
One pound of water =	27.72 cubic inches
	8.33 Gallons
One Gallon of Water =	0.1337 Cubic Feet
	4 Quarts
	8 Pints
	16 Cups

Water Capacities
Average System Water Content US Gallons

Heating Element	Estimated Volume
Cast Iron Radiation	
Radiator, Large Tube	0.114 gal/ sq foot
Radiator, Thin Tube	0.056 gal/ sq foot
Convectors	1.5 Gal/10,000 Btu/Hr @ 200^{0}F
Baseboard	4.7 Gal/10,000 Btu/Hr @ 200^{0}F
Radiation Non Ferrous	
Convectors	0.64 Gal/10,000 Btu/Hr @ 200^{0}F
Baseboard ¾"	.37 Gal/10,000 Btu/Hr @ 200^{0}F
Fan Coil / Unit Htr.	.2 Gal/10,000 Btu/Hr @ 180^{0}F

Water Capacity Steel & Wrought Iron Pipe Schedule 40 US Gallons per Foot			
Pipe Size Inches	Water Capacity/ ft	Pipe Size Inches	Water Capacity/Ft
½"	0.016	3"	0.390
¾"	0.023	4"	0.690
1"	0.040	5"	1.100
1 ¼"	0.063	6"	1.500
1 ½ "	0.102	8"	2.599
2"	0.170	10"	4.096
2 ½"	0.275	12"	5.815

	Water Capacity per Foot in Gallons Copper Tubing		
Pipe Size	Type K	Type L	Type M
3/8"	0.006	0.007	0.008
½"	0.011	0.012	0.013
5/8"	0.017	0.017	
¾"	0.023	0.025	0.027
1"	0.040	0.043	0.045
1 ¼"	0.063	0.065	0.068
1 ½"	0.089	0.092	0.095
2"	0.159	0.161	0.165
2 ½"	0.242	0.248	0.254
3"	0.345	0.354	0.363
4"	.608	.571	.634

WATER

To convert PSIG to feet of water, multiply PSIG x 2.307
One gallon water weighs 8.33 lbs
One cubic foot of water = 7.5 gallons
One cubic foot of water = 1,728 Cubic Inches
One cubic foot of water = 62.4 Pounds
One Pd of Water = 27.72 Cu Inches@65 Deg F
To estimate static pressure in system, multiply highest riser by 0.43 to get pressure at lowest point of system. Always add 4 pounds to get the right pressure for the building.
To estimate pump horsepower required= Horsepower = (GPM x Total head in feet) / 3960
To estimate flow rate of water through a pipe in gallons per minute = GPM = 0.0408 x (pipe diameter)^2 x (water velocity)
To estimate weight of water in a given section of pipe in pounds = Lbs of water = 0.34 x pipe length(feet) x (pipe diameter)^2
Maximum water velocity in pipes should be less than 6 feet per second @ 200^{0}F

Pressure Conversion Chart
Inches H^{2}O to PSI 28" W.C. = 1 psi

Inches H^{2}O	PSI	Inches H^{2}O	PSI
0.1	0.0036	15	0.5414
0.2	0.0072	16	0.5774
0.4	0.0144	17	0.6136
0.6	0.0216	18	0.6496
0.8	0.0289	19	0.6857
1	0.0361	20	0.7218
2	0.0722	21	0.7579
3	0.1083	22	0.7940
4	0.1444	23	0.8301
5	0.1804	24	0.8662
6	0.2165	25	0.9023
7	0.2526	26	0.9384
8	0.2887	27	0.9745
9	0.3248	28	1.010
10	0.3609	29	1.047
11	0.3970	30	1.083
12	0.4331	31	1.191
13	0.4692	32	1.155
14	0.5053	33	1.191

Water Conversion Factors

US Gallons	X	8.34	=	Pounds
US Gallons	X	0.1338	=	Cubic Feet
US Gallons	X	231	=	Cubic Inches
Cu Inch water	X	0.03613	=	Pounds
Cu Inch water	X	0.004329	=	US Gallons
Cu Inch water	X	0.576384	=	Ounces
Pounds Water	X	27.72	=	Cu Inches
Pounds Water	X	0.12	=	US Gallons

Water Pressure to Feet Head

Pounds Per Sq Inch	Feet Head	Pounds Per Sq Inch	Feet Head
1	2.31	100	230.90
2	4.62	110	253.98
3	6.93	120	277.07
4	9.24	130	300.16
5	11.54	140	323.25
6	13.85	150	346.34
7	16.16	160	369.43
8	18.47	170	392.52
9	20.78	180	415.61
10	23.09	200	461.78
15	34.63	250	577.24
20	46.18	300	692.69
25	57.72	350	808.13
30	69.27	400	922.58
40	92.36	500	1154.48
50	115.45	600	1385.39
60	138.54	700	1616.3
70	161.63	800	1847.2
80	184.72	900	2078.1
90	207.81	1000	2309.00

Feet Head to Water Pressure			
Feet Head	Pounds Per Sq Inch	Feet Head	Pounds Per Sq Inch
1	.43	100	43.31
2	.87	110	47.64
3	1.30	120	51.97
4	1.73	130	56.30
5	2.17	140	60.63
6	2.60	150	64.96
7	3.03	160	69.29
8	3.46	170	73.63
9	3.90	180	77.96
10	4.33	200	86.62
15	6.50	250	108.27
20	8.66	300	129.93
25	10.83	350	151.58
30	12.99	400	173.24
40	17.32	500	216.55
50	21.65	600	259.85
60	25.99	700	303.16
70	30.32	800	346.47
80	34.65	900	389.78
90	38.98	1000	433.00

Typical Ground Water Temperatures

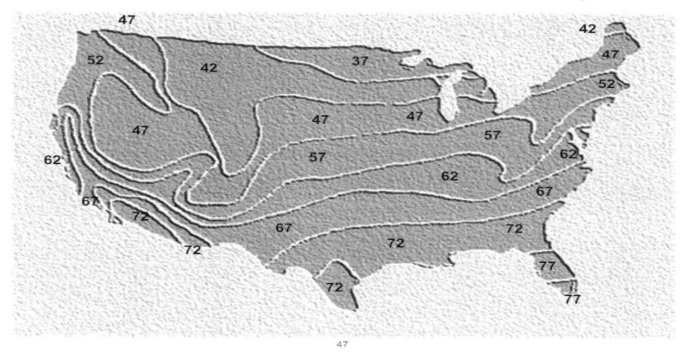

Sizing an Expansion Tank

Recommended Sizing for Expansion Tanks

Nominal Capacity Gallons	Sq ft Radiation
18	350
21	450
24	650
30	900
35	900
35	1100

A rule of thumb for sizing an expansion tank is One gallon for each 23 square feet of radiation or One gallon for each 3,500 Btu of radiation.

If you are going to size an expansion tank, several of the manufacturers have on line calculators that will help you. If you still wish to do a manual calculation, here are the following calculations:

Expansion Tank Sizing
Closed Tank

$$Vt = Vs \frac{[(V2/V1) - 1] - 3\alpha\Delta t}{(P\alpha/P1) - (P\alpha/P2)}$$

Diaphragm Tank

$$Vt = Vs \frac{[(V2/V1) - 1] - 3\alpha\Delta t}{1 - (P1/P2)}$$

Definitions:
Vt = Volume of expansion tank in gallons
Vs = Volume of water in system in gallons
$\triangle T = T_2 - T_1 \, {}^0F$
T_1 = Lower system temperature, typically 40-50^{0}F at fill condition.
T_2 = Higher system design temperature, typically 180^0- 220^{0}F.
$P\alpha$ = Atmospheric pressure (14.7 Psia)
P_1 = System fill pressure Minimum System pressure (Psia)
P_2 = System operating pressure Maximum System pressure (Psia)

α = Linear Coefficient of expansion
Steel 6.5 x 10^{-6}
Copper 9.5 x 10^{-6}

When choosing a diaphragm tank, use the acceptance factor when choosing the size. The acceptance factor is the amount of space that is available in the tank.

To see how this formula works, let us look at a hypothetical building. Our building has a system volume of 2,000 gallons. The system will operate between 180^{0}F and 220^{0}F. The minimum pressure will be 10# and the maximum pressure will be at 25#. 14.7 has to be added to each pressure to get atmospheric pressure. For example, the low pressure will be 14.7# plus 10# = 24.7#. The high pressure will be 14.7# plus 25# = 39.7#. Our relief valve is set at 30#. The system is steel pipe. The volume of the water is as follows:
V_1 = 40^{0}F = 0.01602 ft^2/lb (Ground water temperature)
V_2 = 220^{0}F = 0.01677 ft^2/lb (Design Temperature)
α = coefficient of thermal expansion for steel pipe is 6.5 x 10^{-6}

We will size a closed tank. Here is the formula again.

$$Vt = Vs \frac{[(V2/V1) - 1] - 3\alpha\Delta t}{(P\alpha/P1) - (P\alpha/P2)}$$

Vt
$$= 2000 \frac{[(0.01677/0.01602) - 1] - 3(6.5 \times 0.000001) \times 180)}{(14.7/24.7) - (14.7/39.7)}$$

$$Vt = 2000 \frac{0.0468 - 0.00351}{0.595 - 0.370}$$

$$Vt = 2000 \frac{0.0433}{.225}$$

Vt = 385 Gallons

Estimated system water volume 35 Gallons per Boiler HP
Typical system fill pressure 10 Psi

139

Rule of Thumb for Expansion Tank Sizing Steel Piping

Based upon the following:
Entering Pressure 10#, Maximum Pressure 25#
Entering Temperature 40°F, Maximum Temperature 220°F

Steel Piping		
System Capacity in Gallons	Closed Expansion Tank	Diaphragm Tank
200	39	23
300	58	34
400	77	46
500	96	57
600	116	69
700	135	80
800	154	92
900	173	103
1000	193	115
1100	212	126
1200	231	138
1300	250	149
1400	270	160
1500	289	172
1600	308	183
1700	327	195
1800	347	206
1900	366	218
2000	385	229
2500	481	287
3000	578	344
3500	674	401
4000	770	458
4500	867	516
5000	963	573
6000	1156	688
7000	1348	802
8000	1541	917
9000	1733	1032
10000	1926	1146

Rule of Thumb for Expansion Tank Sizing Copper Piping

Based upon the following:
Entering Pressure 10#, Maximum Pressure 25#
Entering Temperature 40°F, Maximum Temperature 220°F

Copper Piping		
System Capacity in Gallons	Closed Expansion Tank	Diaphragm Tank
200	37	22
300	56	33
400	74	44
500	93	55
600	111	66
700	130	77
800	148	88
900	167	99
1000	185	110
1100	204	121
1200	222	132
1300	241	143
1400	260	154
1500	278	165
1600	297	177
1700	315	188
1800	334	199
1900	352	210
2000	371	221
2500	463	276
3000	556	331
3500	649	386
4000	742	441
4500	834	496
5000	927	552
6000	1112	662
7000	1298	772
8000	1483	883
9000	1668	993
10000	1854	1103

How to Estimate Hydronic System Volume

In many instances, you will need to calculate system water volume. This is useful when estimating water treatment and or glycol requirements. The following are some ideas that may help you to do that. The most accurate method is to measure and note the actual pipe sizes in the hydronic loop. This could be done by consulting the building blueprints. This is the most accurate method. There are several other rules of thumb that are used in the industry. I have a list of these below. A rule of thumb is that a hot water loop will be about 2/3 the size of a chilled water loop.

Rules of Thumb to Estimate System Volume

- Multiply steel expansion tank volume by 5.
- 35 – 50 gallons per Boiler HP
- Pump GPM x 4
- Expansion Tank volume is 20% of system volume
- Rated tonnage of system x 10 gallons

The Salt Test

A common method for estimating system volume is to use salt because it is easy to test for, very soluble, and inexpensive. The disadvantage to this type of test is that the system has to be flushed at the end of the test, wasting water and chemicals. If not, the high chloride levels can be corrosive to the system.

Salt Test Procedure

1. Fill The system with fresh water. Circulate and flush the system until the water is clear. Eliminate all sources of water loss such as bleed, overflow, etc.

2. Measure the chloride Cl concentration in the system and estimate the system volume.

3. Add one pound of Table Salt (Sodium Chloride) per 1,000 gallons of estimated volume. This can be added in the pot feeder. Verify that the salt mixes thoroughly.

4. Allow one hour for the salt to be mixed into the system.

5. Re-Measure the chloride concentration

6. Multiply the estimated gallons of water by 76 ppm. Divide this by the difference (increase) in chloride concentration.

7. The answer will be the actual system volume

Example

Estimated Volume	1,000 Gallons
Initial Chloride Test	100 ppm
Final Chloride Test	180 ppm

Calculation

$$\frac{1,000 \; gals. \times 76 \; ppm}{(180-100) ppm} = 950 \; \text{Actual Gallons in Loop}$$

Flush system rapidly to return the chloride level to normal.

Misc Piping Information
PIPING EXPANSION

Calculating the linear expansion of pipe carrying steam or hot water

If you would like to calculate the expansion or lengthening of a pipe when it has steam or hot water inside, try this formula

E= constant x (T-F)
E = Expansion in inches per 100 feet of pipe
F = Starting temperature
T = Final temperature

Constant = Coefficient of expansion per 100 Ft pipe

Metal	Constant
Steel	0.00804
Wrought Iron	0.00816
Cast Iron	0.00780
Copper or Brass	0.01140

For Example,

What is the expansion of 100 feet of steel pipe that carries 10 # steam pressure?
E=constant x (T-F)
E= 0.00804 x (239 (steam temperature) – 50(starting temp)
E= 0.00804 x 189
E = 1.51" expansion per 100'

Heat Losses from Horizontal Steel Pipe
Btu per hour per linear foot at 70 Deg E room temperature

Nom Pipe Size	Hot Water (180 Deg F)	Steam 5 PSIG
½	60	96
¾	73	118
1	90	144
1 ¼	112	170
1 ½	126	202
2	155	248
2 ½	185	296
3	221	355
4	279	448

Heat Losses from Horizontal Copper Pipe
Btu per hour per linear foot at 70 Deg E room temperature

Nom Pipe Size	Hot Water (180 Deg F)
½	33
¾	45
1	55
1 ¼	66
1 ½	77
2	97
2 ½	117
3	136
4	174

Thermal Expansion of Piping Material in inches per 100 feet above 32 Deg F

Temperature Deg F	Carbon & Carbon Moly Steel	Cast Iron	Copper
32	0	0	0
100	0.5	0.5	0.8
150	0.8	0.8	1.4
200	1.2	1.2	2.0
250	1.7	1.5	2.7
300	2.0	1.9	3.3
350	2.5	2.3	4.0
400	2.9	2.7	4.7
500	3.8	3.5	6.0
600	4.8	4.4	7.4
700	5.9	5.3	9.0

Avg. Heat Loss from Insulated Pipe Btuh/Linear Feet @ 70°F			
Pipe Size Inches	Insulation Thickness	175^0F	225^0F
½	1"	0.150	0.157
¾	1"	0.172	0.177
1	1"	0.195	0.200
	1 ½"	0.165	0.167
1 ¼"	1"	0.250	0.232
	1 ½"	0.170	0.193
1 ½"	1"	0.247	0.255
	1 ½"	0.205	0.210
2"	1"	0.290	0.297
	1 ½"	0.235	0.240
	2"	0.200	0.205
2 ½"	1"	0.330	0.340
	1 ½"	0.265	0.270
	2"	0.225	0.230
3"	1"	0.385	0.395
	1 ½"	0.305	0.312
	2"	0.257	0.263
4"	1"	0.470	0.480
	1 ½"	0.370	0.379
	2"	0.308	0.315

Standard Copper Tubing Type K,L,M		
Pipe Size	Outside Diameter (O.D.)	Circumference
½"	0.625"	1.964"
¾"	0.875"	2.749"
1"	1.125"	3.534"
1 ¼"	1.375"	4.319"
1 ½"	1.625"	5.105"
2"	2.125"	6.675"
2 1/2"	2.625"	8.246"
3"	3.125"	9.817"
4"	4.125"	12.959"
6"	6.127"	12.248"
8"	8.125"	25.525"
10"	10.125"	31.808"
12"	12.750"	40.054"

Standard Nipples & Pipe Sizing Schedule 40		
Pipe Size	Outside Diameter (O.D.)	Circumference
1/8"	0.405"	1.272"
¼ "	0.540"	1.696"
3/8"	0.675"	2.121"
½"	0.840'	2.639"
¾"	1.050"	3.299"
1"	1.315"	4.131"
1 ¼"	1.660"	5.215"
1 ½"	1.900"	5.969"
2"	2.375"	7.461"
2 1/2"	2.875"	9.032"
3"	3.500"	10.995"
4"	4.500"	14.137"
5"	5.563"	17.476"
6"	6.625"	20.812"
8"	8.625"	27.095"
10"	10.750"	33.771"
12"	12.750"	40.054"

Frictional Allowance for Fittings In Feet of Pipe					
	Length to be added in feet				
Size of Fittings Inches	90^0 Ell	Side Outlet Tee	Gate Valve	Globe Valve	Angle valve
1/2"	1.3	3	0.3	14	7
3/4'	1.8	4	0.4	18	10
1"	2.2	5	0.5	23	12
1 1/4"	3.0	6	0.6	29	15
1 1/2"	3.5	7	0.8	34	18
2"	4.3	8	1.0	46	22
2 1/2"	5.0	11	1.1	54	27
4"	9	18	1.9	92	45
6"	13	27	2.8	136	67
8"	17	35	3.7	180	92
10"	21	45	4.6	230	112

Fuel & Flue Information & Piping

Flue Information

Diameter	Circumference	Diameter	Circumference
12	37.70	28	87.96
14	43.98	30	94.25
16	50.27	32	100.53
18	56.55	34	106.81
20	62.83	36	113.10
22	69.12	38	119.38
24	75.40	40	125.66
26	81.68		

Typical Boiler Exhaust Velocity	
Equipment Type	**Typical Exhaust Velocity ft/s**
Boiler with On Off Burner	16-26
Boiler with two step burner	31-49
Boiler with modulating burner	49-82
Minimum to keep surface free from soot	9.8-13

Flue Gas Condensing Temperatures	
Fuel	**Condensing Temperature**
Natural Gas	250 ^{0}F
#2 Oil	275 ^{0}F
#6 Oil	300 ^{0}F
Coal	325 ^{0}F
Wood	400 ^{0}F

Comparative Fuel Values
To get 1,000,000 Btu's you need the following

Fuel Source	1,000,000 Btu's
Natural Gas @ 1000 Btu/ cu ft	1,000 Cu ft
Coal @ 12,000 Btu/ lb	83.333 Lb
Propane @ 91,600 Btu/ gal	10.917 Gal
Gasoline @ 125,000 btus/gal	8.000 Gal
Fuel Oil #2 @ 140,000 btus/gal	7.194 Gal
Fuel Oil #6 @ 150,000 btus/gal	6.666 Gal
Electricity @ 3,412 Btu/kWh	293.083 Kwh

Average Btu Content of Common Fuels

Fuel Type	Number of Btu/ Unit
Fuel Oil #2	140,000 / Gallon
Natural Gas	1,025,000/ 1,000 cubic feet
Propane	91,330/ gallon
Coal	28,000,000 per ton
Electricity	3,412/ KWH
Wood (air dried)	20,000,000/ cord or 8,000 / pound
Kerosene	135,000 / gallon
Pellets	16,500,000/ton

Misc Fuel Information

Natural Gas	
1 Cu ft Natural Gas	1,000 Btus
1 MCF =	1,000,000 Btus
	1,000 Cu. Ft.
	10 CCF
	10 Therms
1 Dekatherm	1 MCF
	10 Therms
	1,000,000 Btus
1 Therm =	100,000 Btus
	0.1 MCF
	100 Cu. Feet
1 CCF =	1,000 Cu Ft
	100 Therm
Propane	
1 gallon	92,000 Btus
1 Cu Foot	2,250 Btus
#2 Fuel Oil	
1 Gallon	140,000 Btus

Gas Pipe Line Sizing

Steel Pipe Size	Pipe Length			
	10 Feet	20 Feet	40 feet	80 Feet
	Capacity in Cubic Feet per hour			
½"	120	85	60	42
¾"	272	192	136	96
1"	547	387	273	193
1 1/4"	1,200	849	600	424
1 ½"	1,860	1,316	930	658
2"	3,759	2,658	1,880	1,330
2 ½"	6,169	4,362	3,084	2,189
4"	23,479	16,602	11,740	8,301

Each cubic foot of gas roughly equals 1,000 Btuh

CSST Sizing
Rules of Thumb for EHD Sizing

EHD	Pipe Size	EHD	Pipe Size
15	3/8"	37	1 1/4"
19	1/2"	46	1 1/2"
25	3/4"	62	2"
31	1"		

Please check with the manufacturer to verify their sizing

Sizing Corrugated Stainless Steel Tubing (CSST) Natural Gas

Inlet Pressure	Pressure Drop	Specific Gravity
Less than 2 psi	0.6"	0.60

Tube Size	Length				
EHD	5	10	15	20	25
	Capacity in Cu ft per Hour				
13	46	32	25	22	19
15	63	44	35	31	27
18	115	82	66	58	52
19	134	95	77	67	60
23	225	161	132	116	104
25	270	192	157	137	122
30	471	330	267	231	206
31	546	383	310	269	240
37	895	639	524	456	409
46	1792	1260	1030	888	793
48	2070	1470	1200	1050	936
60	3660	2930	2400	2080	1860
EHD	30	40	50	60	70
	Capacity in Cu ft per Hour				
13	18	15	13	12	11
15	25	21	19	17	16
18	47	41	37	34	31
19	55	47	42	38	36
23	96	83	75	68	63
25	112	97	87	80	74
30	188	162	144	131	121
31	218	188	168	153	141
37	374	325	292	267	248
46	723	625	559	509	471
48	856	742	665	608	563
60	1520	1320	1180	1080	1000

EHD = Equivalent Hydraulic Diameter
Each cubic foot of gas equals 1,000 Btuh

Gas Pressure Comparison

Inches Hg	Ounces	PSI	Inches Hg	Ounces	PSI
0.1	0.05	0.003	15	8.67	0.54
0.2	0.12	0.007	16	9.24	0.576
0.4	0.23	0.01	17	9.82	0.612
0.6	0.35	0.02	18	10.4	0.648
0.8	0.46	0.028	19	10.98	0.684
1	0.58	0.036	20	11.56	0.72
2	1.15	0.072	21	12.13	0.756
3	1.73	0.108	22	12.71	0.792
4	2.31	0.144	23	13.29	0.828
5	2.89	0.18	24	13.87	0.864
6	3.46	0.216	25	14.45	0.9
7	4.04	0.252	26	15.02	0.936
8	4.62	0.288	27	15.60	0.972
9	5.20	0.324	28	16.18	1.008
10	5.78	0.36	29	16.76	1.044
11	6.35	0.396	30	17.34	1.08
12	6.93	0.432	31	17.91	1.16
13	7.51	0.468	32	18.49	1.152
14	8.09	0.504	33	19.07	1.188

Orifice Capacities for Natural Gas

1000 Btu per cubic foot

Manifold pressure 3 ½" Water Column

Wire Gauge Drill Size	Rate Cu Ft / Hr	Rate Btu/Hr
70	1.34	1,340
68	1.65	1,650
66	1.80	1,870
64	2.22	2,250
62	2.45	2,540
60	2.75	2,750
58	3.50	3,050
56	3.69	3,695
54	5.13	5,125
52	6.92	6,925
50	8.35	8,350
48	9.87	9,875
46	11.25	11,250
44	12.62	12,625
42	15.00	15,000
40	16.55	16,550
38	17.70	17,700
36	19.50	19,500
34	21.05	12,050
32	23.70	23,075
30	28.50	28,500
28	34.12	34,125
26	37.25	37,250
24	38.75	38,750
22	42.50	42,500
20	44.75	44,750

How Long Will It Last?

Source ASHRAE

Equipment	Years
Boilers	24-35
Burners	21
Rooftop Units	15
Furnaces	18
Pumps, Base Mounted	20
Pumps Pipe Mounted	10
Pumps, Condensate	15
Condensate Piping	10-20
Steam Traps	7 *based upon US GSA*

Misc Boiler Information

1 hp = 0.746 KW = 746 WATTS = 2,545 BTUH = 1.0KVA
1 KW = 1,000 WATTS = 3,413 BTUH = 1.341 HP
1 WATT = 3.415 BTUH
1 BTUH = Btu per Hour
1 Boiler HP = 34,500 Btuh = 34.5 Lbs Steam/ Hr = 34.5 Lb H2O/ Hr = 0.069 GPM = 4.14 GPH = 140 EDR
1lb Steam = 0.002 GPM
1 EDR = 0.000496 gpm = 0.25 lbs steam cond/hr
1000 EDR = 0.496 gpm
1 EDR hot water = 150 Btuh
1 EDR steam = 240 Btuh
1 Btu will raise 1 cubic feet of air 55 degrees
1 Btu will raise 55 cubic feet of air 1 degree F
1 Btu = Amount of heat required to raise one pound of water one degree.
One Boiler HP requires about 10-11 ½ square feet of boiler heating surface.
One Boiler HP = 34.5 lbs of steam/ hr from and at 212 degrees F.
The products of combustion produced when 1 cubic foot of gas is completely burned are 8 cubic feet of nitrogen, 1 cubic foot of carbon dioxide and 2 cubic feet of water vapor.
Average heating loads 25-40 Btuh/ square foot

Common Boiler Calculations	
To Find?	Perform this calculation
Lbs Steam per Hour	Boiler HP x 34.5
Evaporation Rate GPM	Boiler HP x .069
MBTU per Hr Output (MBH)	Boiler HP x 33.4
Sq. Ft of Equivalent Direct Radiation (EDR)	Boiler HP x 139
Sq Ft Equivalent Direct Radiation (EDR)	BTU/Hr /240
Evaporation Rate GPM	EDR/1000 x 0.5
Evaporation rate GPM	Lbs Steam Hr / 500
BTU	500 x GPM x Delta T
GPM	BTU / 500 x Delta T
Delta T	BTU /500 x GPM
Calculate Sensible Heat	1.08 x CFM x Temp Rise or Delta T
Calculate Sensible Heat	500 x GPM x Temp Rise or delta T

1 W =	0.00134 hp
	3.414 Btu
	0.0035 lb of water evaporated per hour
1 kW =	1,000 W
	1.34 hp
	3.53 lbs water evaporated per hr from and at 212 degrees F
	0.955 Btu's
	57.3 BTUm
	3415 Btuh
1 HP =	746 W
	0.746 kW
	33,000 ft-lb per minute
	550 ft-lb per second
	33,475 Btuh
	34.5 Lbs Steam/hr from and at 212Degrees F
	42.746 Btu/min
	2564.76 Btuh
1 Kwh =	1,000 W/Hr
	1.34 hp/hr
	3,600,000 joules
	3.53 lbs water evaporated per hr from and at 212 degrees F
	22.75 lbs water raised from 62 degrees F to 212 degrees F
1 Btu =	17.452 W/Min
	0.2909W/Hr

ELECTRICAL
Amp of Copper Wire Types
Single wire in open air

Wire Size AWG	TH UF	FEPW, RH, RHW, TWH, THWN, ZW, THHW, XHHW	USE-2, XHH, XHHW, TBS, SA, SIS, FEP, MI, RHW-2, THHN, ZW-2, THWN-2, FEPB, RHH, THHW, THW-2
0000	300	360	405
000	260	310	350
00	225	265	300
0	195	230	260
1	165	195	220
2	140	170	190
3	120	145	165
4	105	125	140
6	80	95	105
8	60	70	80
10	40	50	55
12	30	35	40
14	25	30	35
16	-	-	24
18	-	-	18

Up to 86-degree ambient temperature

Amp of Copper Wire Types
Three wires in cable

Wire Size AWG	TH UF	FEPW, RH, RHW, TWH, THWN, ZW, THHW, XHHW	USE-2, XHH, XHHW, TBS, SA, SIS, FEP, MI, RHW-2, THHN, ZW-2, THWN-2, FEPB, RHH, THHW, THW-2
0000	195	230	260
000	165	200	225
00	145	175	195
0	125	150	170
1	110	130	150
2	95	115	130
3	85	100	110
4	70	85	95
6	55	65	75
8	40	50	55
10	30	35	40
12	25	25	30
14	20	20	25
16	-	-	18
18	-	-	14

Up to 86-degree ambient temperature

Electrical Formulas
Current = Amps = I
Volts = Voltage = E
Resistance = Ohms = R
Watts = W

Ohms Law

$$\text{Current} = \frac{Voltage}{Resistance} \text{ or } I = \frac{E}{R}$$

$$R = \frac{E}{I} \qquad \text{or} \qquad E = I \times R$$

Power
Watts = Volts x Amps or W=E x I

$$E = \frac{W}{I} \qquad \text{or} \qquad I = \frac{W}{E}$$

Standard 24 Volt Thermostat Connections		
Terminal	Usage	Normal Colors
R or V	24 VAC power	Red
Rh or 4	24 VAC Heating Power	Red
Rc	24 VAC Cooling Power	Red
C	24 VAC Common	Black
Y	1st Stage Cooling	Yellow
Y2	2nd Stage Cooling	Blue or Orange
W	1st Stage Heat	White
W2	2nd Stage Heat	No Standard Color
G	Fan	Green

Full Load Amperes of Single Phase Motors			
HP	RPM	115V	230V
1/8	1725	2.8	1.4
	1140	3.4	1.7
	860	4.0	2.0
1/4	1725	4.6	2.3
	1140	6.15	3.07
	860	7.5	3.75
1/3	1725	5.2	2.6
	1140	6.25	3.13
	860	7.35	3.67
1/2	1725	7.4	3.7
	1140	9.15	4.57
	860	12.8	6.4
3/4	1725	10.2	5.1
	1140	12.5	6.25
	860	15.1	7.55
1	1725	13.0	6.5
	1140	15.1	7.55
	860	15.9	7.95

Full Load Amperes of Three Phase Motors			
HP	RPM	115V	230V
1/4	1725	0.95	0.48
	1140	1.4	0.7
	860	1.6	0.8
1/3	1725	1.19	0.6
	1140	1.59	0.8
	860	1.8	0.9
1/2	1725	1.72	0.86
	1140	2.15	1.08
	860	2.38	1.19
3/4	1725	2.46	1.23
	1140	2.92	1.46
	860	3.26	1.63
1	1725	3.19	1.6
	1140	3.7	1.85
	860	4.12	2.06
1 1/2	1725	4.61	2.31
	1140	5.18	2.59
	860	5.75	2.88
2	1725	5.98	2.99
	1140	6.5	3.25
	860	7.28	3.64
3	1725	8.70	4.35
	1140	9.25	4.62
	860	10.3	5.15
5	1725	14.0	7.0
	1140	14.6	7.3
	860	16.2	8.1
7 1/2	1725	20.3	10.2
	1140	20.9	10.5
	860	23.0	11.5

Calculate Motor HP from Meter Readings

DC Motors

$$HP = \frac{volts \times amperes \times efficiency}{746}$$

Single Phase AC Motors

$$HP = \frac{volts \times amperes \times efficiency \times power\ factor}{746}$$

Three Phase AC Motors

$$HP = \frac{volts \times amperes \times efficiency \times power\ factor \times 1.73}{746}$$

Typical Flame Safeguard Signals Siemens	
Gas Burner Controls	
Model	**Flame Signal**
LFL with UV sensor QRA Minimum 70 uA DC	100-450 µA DC Typical
LFL with Flame Rod Minimum 6 uA DC	20-100 uA DC
Oil Burner Controls	
Model	**Flame Signal**
LAL1 with photoresistive detector, QRB1	95-160 µA DC
LAL1 with blue-flame detector, QRC1	80-130µA DC
LAL2/LAL3 with photoresistive detector, QRB1	8-35 µA DC
LAL2/LAl3 with selenium photocell detector, RAR	6.5-30 µA DC
LAL4 with photoresistive detector, QRB1	95-160 µA DC
LAL4 with blue-flame detector, QRC1	80-130 µA DC

Flame Safeguard Definitions	
µA	Micro Amps
VDC	Volts DC

Typical Flame Safeguard Signals	
Fireye	
Model	**Average Flame Signal**
UVM	4.0-5.5 VDC
TFM	14-17 VDC
D-10/20/30	16-25 VDC
E-100/ 110	20-80 VDC
E-100/E110 with EPD Programmer	4-10 VDC
M Series II	4-6 VDC
Micro M Series	4-10 VDC
Micro M Series w Display	20-80 VDC
Honeywell	
Model	**Average Flame Signal**
RA890	2-6 µA DC
R4795	2-6 µA DC
R7795	2-6 µA DC
R4140	2-6 µA DC
R4150	2-6 µA DC
BC7000	2-6 µA DC
RM7890	1.25-5 VDC
RM7895	1.25 VDC
RM7840	1.25 VDC
RM7800	1.25 VDC

Estimate Storage Tank Capacity in Gallons
Rectangular Tank
Sizing a Storage Tank

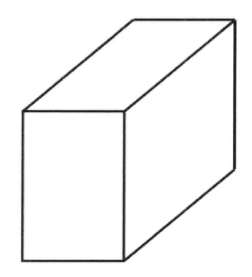

Rectangular Storage Tanks

Length" x Width" x Height" divided by 231 = Gallons

Measure tank length, width and height in inches
Multiply Length x Width x Length
Divide that total by 231 = Gallons
The following are some tank sizes that are already calculated.

Rectangular Tank Height 18"				
Length	**Width**			
	US Gallons			
	12	**18**	**24**	**36**
12	11	17	22	34
18	17	25	34	50
24	22	34	45	67
30	28	42	56	84
36	34	50	67	101
42	39	59	79	118
48	45	67	90	135
60	56	84	112	168

Rectangular Tank Height 24"				
Length	**Width**			
	US Gallons			
	12	**18**	**24**	**36**
12	15	22	30	45
18	22	34	45	67
24	30	45	60	90
30	37	56	75	112
36	45	67	90	135
42	52	79	105	157
48	60	90	120	180
60	75	112	150	224

Rectangular Tank Height 12"				
Length	**Width**			
	US Gallons			
	12	**18**	**24**	**36**
12	7	11	15	22
18	11	17	22	34
24	15	22	30	45
30	19	28	37	56
36	22	34	45	67
42	26	39	52	79
48	30	45	60	90
60	37	56	75	112

Rectangular Tank Height 36"				
Length	**Width**			
	US Gallons			
	12	**18**	**24**	**36**
12	22	34	45	67
18	34	50	67	101
24	45	67	90	135
30	56	84	112	168
36	67	101	135	202
42	79	118	157	236
48	90	135	180	269
60	112	168	224	337

Estimate Storage Tank Capacity in Gallons

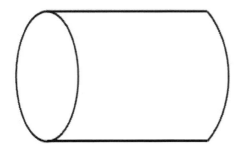

Round Storage Tanks
Multiply ½ Tank Diameter by itself
Multiply that by 3.146 x Length of Tank in inches
Divide by 231 = Gallons of water
For example, you have a 24" diameter tank that is 36" long
½ of 24" = 12" x 12" = 144
144 x 3.146 x 36 = 16,308
16,308 divided by 231 = 70.59 Gallons

Length (feet)	Circular Tank Inside Diameter (Inches)				
	US Gallons				
	18	24	30	36	42
1	1.1	1.96	3.06	4.41	5.99
2	26	47	73	105	144
2.5	33	59	91	131	180
3	40	71	100	158	216
3.5	46	83	129	184	252
4	53	95	147	210	288
4.5	59	107	165	238	324
5	66	119	181	264	360
5.5	73	130	201	290	396
6	79	141	219	315	432
6.5	88	155	236	340	468
7	92	165	255	368	504
7.5	99	179	278	396	540
8	106	190	291	423	576
9	119	212	330	476	648
10	132	236	366	529	720
12	157	282	440	634	864
14	185	329	514	740	1008

Circular Tank Inside Diameter (Inches)					
US Gallons					
Length (feet)	48	54	60	66	72
1	7.83	9.91	12.24	14.41	17.62
2	188	238	294	356	423
2.5	235	298	367	445	530
3	282	357	440	534	635
3.5	329	416	513	623	740
4	376	475	586	712	846
4.5	423	534	660	800	852
5	470	596	734	899	1057
5.5	517	655	808	978	1163
6	564	714	880	1066	1268
6.5	611	770	954	1156	1374
7	658	832	1028	1244	1480
7.5	705	889	1101	1355	1586
8	752	949	1175	1424	1691
9	846	1071	1322	1599	1903
10	940	1189	1463	1780	2114
12	1128	1428	1762	2133	2537
14	1316	1666	2056	2490	2960

MISC

Pressure Unit Conversions

Known	Desired Pressure Unit			
	Pounds Per sq In.	Ounces Per Sq In.	Inches of Water	Feet of Water
Centimeters of Water	0.0981	0.227	0.384	0.0328
Feet of water	0.433	6.94	12.0	0.883
Inches Mercury	0.491	7.86	13.6	1.13
Inches Water	0.0361	0.578	-------	0.0833
Ounces per Square Inch	0.0625	---------	1.73	0.144
Pounds per Sq Inch	----------	16.0	27.7	2.31

Common Fraction to Decimal to Millimeters		
Fraction	Decimal	Millimeters
1/16	0.0625	1.587
1/8	0.125	3.175
3/16	0.1875	4.762
1/4	0.250	6.350
5/16	0.3125	7.937
3/8	0.375	9.525
7/16	0.4375	11.113
1/2	0.50	12.700
9/16	0.5625	0.5625
5/8	0.625	15.875
11/16	0.6875	17.462
3/4	0.750	19.050
13/16	0.8125	20.637
7/8	0.875	22.225
15/16	0.9375	23.812
1	1.00	25.400

Metric

Metric Liquid	
Metric	U.S.
3.7854 L	1 Gallon
0.946 L	1 Quart
0.473 L	1 Pint
1 L	0.264 Gallons
1 L	33.814 Ounces
29.576 ml	1 Fluid Ounce
236.584 ml	1 Cup
Metric Length	
Metric	U.S.
1m	39.37 inches
1 m	3.28 feet
1 m	1.094 yards
1 m	0.0016 miles
1.609 km	1 mile
25.4 mm	1 inch
2.54 cm	1 inch
304.8 mm	1 foot
1 mm	0.03937 inches
1 cm	0.3937 inches
1 dm	3.937 inches
Metric Pressure	
6.8947 kPa	1 pound per sq in (psi)
9.794 kPa	1m column of water
1 kPa	10.2 cm of water
1.3332 kPa	1 cm column of water
3.3864 kPa	1 inch of mercury (in Hg)
8 kPa	6 cm of mercury
Metric Conversions	
KJ/Hr =	Btu/h x 1.055
CMM =	CFM x 0.02832
LPM =	GPM x 3.785
Kj/Lb.=	Btu/Lb x 2.326
Meters =	Feet x 0.3048
Sq Meters =	Sq Feet x 0.0929
Cu. Meters =	Cu. Feet x 0.02832
Kg =	Pounds x 0.4536
Kg/Cu. Meter =	Pounds. Cu Feet x 16.017
Cu. Meters/ Kg =	Cu. Ft/ Pound x 0.0624

CONVERSION FACTORS	
1 Mile =	1,760 yards
	5,280 feet
	63,360 inches
	1.609 Km
1 Foot =	0.3048 M
	30.48 Cm
	304.8 mm
1 Inch =	25,400 microns
1 Gallon H2O =	8.333 Lbs
1 Lb =	16 oz
	7,000 grains
	0.4536 Kg
1 ton =	2,000 lbs
	907 Kg
1 lb steam =	1 lb H2O
1 square foot =	144 square inches
1 acre =	43,560 Sq Ft
	4,840 Sq Yds
	0.4047 Hectares
1 Sq Mile =	640 Acres
1 Sq Yd =	9 Sq Ft
	1,296 Sq Inches
1 Sq Foot =	Square yards x 9
1 Cu Yard =	27 Cu Ft
	46,656 cu inches
	1,616 pints
	807.9 quarts
	764.6 Liters
1 Cu foot =	1,728 cubic inches
1 Liter =	0.2642 Gallons
	1.057 quarts
	2.113 pints
1 gallon =	4 quarts
	8 pints
	3.785 liters
	0.13368 Cu Feet
	231 Cu Inches
1 Barrel Oil =	42 gallons
1 MPH =	5280 ft / hr
	88 ft/min
	1.467 ft/sec
	0.868 Knots per Hr
1 Knot =	1.1515 MPH
1 League =	3.0 Miles
Speed of Sound in Air =	1,128.5 ft/sec
	769.4 mph
14.7 psi =	33.95 ft H2O
	29.92 in Hg
	407.2 In wg

	2,116.8 Lbs/ Sq Ft
1 Psia =	Psig = 14.7
1 psi =	2.307 Ft H2O
	2.036 In Hg
	16 ounces
	27.7 In w.c.
1 ounce =	1.73 inches w.c
1 Ft H20 =	0.4335 psi
	62.43 lbs/ sq feet
Diameter of Circle =	Circumference x 0.3188
Circumference of Circle =	Diameter x 3.1416
Hours in a Year	8,760

Trouble Shooting Boilers with Atmospheric Burner

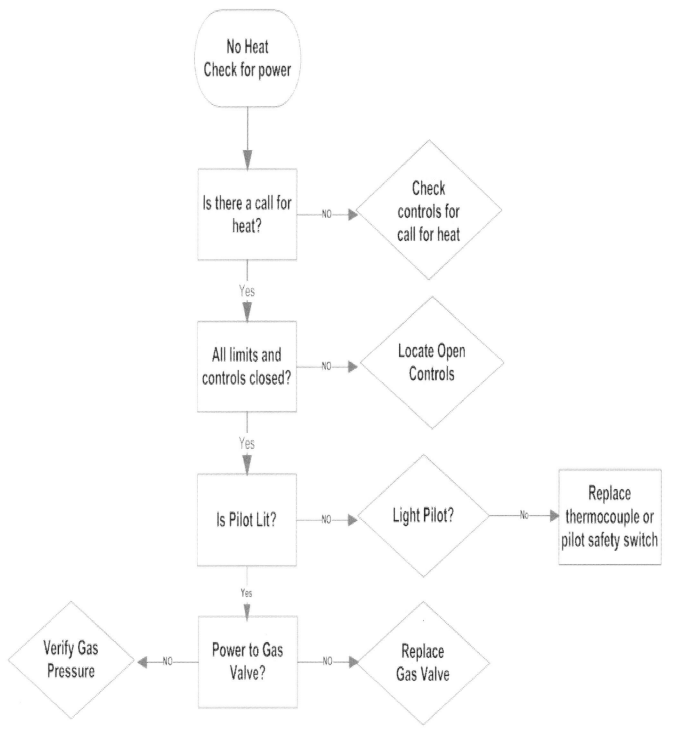

Trouble Shooting Boilers with Power Burner

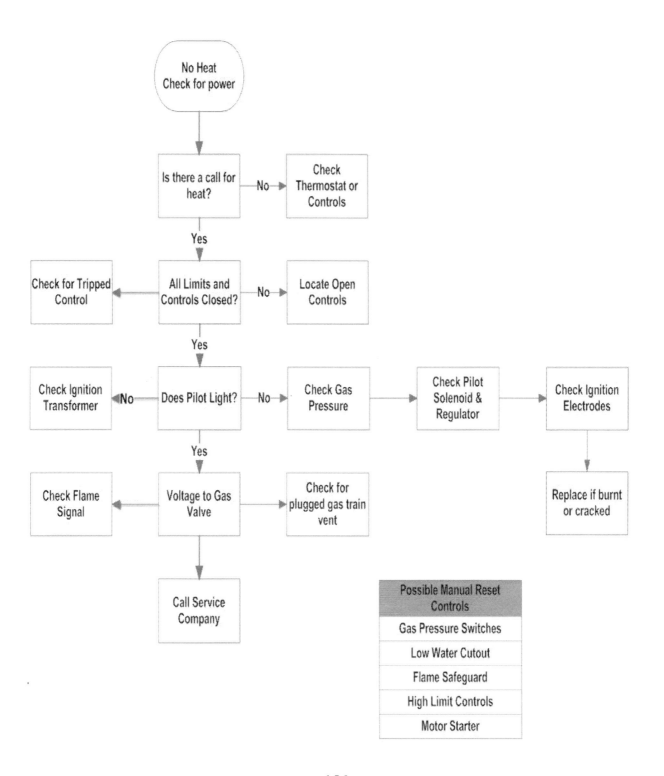

Possible Manual Reset Controls
Gas Pressure Switches
Low Water Cutout
Flame Safeguard
High Limit Controls
Motor Starter

Degree Day Definition

Early this century, heating engineers developed the concept of heating degree-days as a useful index of heating fuel requirements. They found when the daily mean temperature is lower than 65 degrees, most buildings require heat to maintain an inside temperature of 70 degrees. The daily mean temperature is obtained by adding together the maximum and minimum temperatures reported for the day and dividing the total by two. Each degree of mean temperature below 65 is counted as one heating degree-day. Thus, if the maximum temperature is 70 degrees and minimum 52 degrees, four heating degree-days would be produced. (70 + 52 = 122; 122 divided by 2 = 61; 65-61 = 4.) If the daily mean temperature is 65 degrees or higher, the heating degree-day total is zero. Suppliers of heating fuels use this measure to estimate their customers' fuel requirements.

Degree Days and Design Temperatures

State	Station	Heating Degree Days	Winter Design Temperature
AL	Birmingham	2551	21
	Huntsville	3,070	16
	Mobile	1,560	29
	Montgomery	2,291	25
AK	Anchorage	10,864	-18
	Fairbanks	14,279	-47
	Juneau	9,075	1
	Nome	14,171	-27
AZ	Flagstaff	7,152	4
	Phoenix	1,765	34
	Tucson	1,800	32
	Yuma	974	39
AR	Fort Smith	3,292	17
	Little Rock	3,219	20
	Texarkana	2,533	23
CA	Fresno	2,611	30
	Long Beach	1,803	43
	Los Angeles	2,061	43
	Los Angeles	1,349	40
	Oakland	2,870	36
	Sacramento	2,502	32
	San Diego	1,458	44
	San Francisco	3,015	38
	San Francisco	3,001	40

State	Station	Heating Degree Days	Winter Design Temperature
CO	Alamosa	8,529	-16
	Colorado Springs	6,423	2
	Denver	6,283	1
	Grand Junction	5,641	7
	Pueblo	5,462	0
CT	Bridgeport	5,617	9
	Hartford	6,235	7
	New Haven	5,897	7
DE	Wilmington	4,930	14
DC	Washington	4,224	17
FL	Daytona		
	Fort Myers	442	44
	Jacksonville	1,239	32
	Key West	108	57
	Miami	214	47
	Orlando	766	38
	Pensacola	1,463	29
	Tallahassee	1,485	30
	Tampa	683	40
	West Palm Beach	253	45
GA	Athens	2,929	22
	Atlanta	2,961	22
	Augusta	2,397	23
	Columbus	2,383	24
	Macon	2,136	25
	Rome	3,326	22
	Savannah	1,819	27
HI	Hilo	0	62
	Honolulu	0	63
ID	Boise	5,809	10
	Lewiston	5,542	6
	Pocatello	7,033	-1
IL	Chicago (Midway)	6,155	0
	Chicago (O'Hare)	6,639	-4
	Chicago	5,882	2
	Moline	6,408	-4
	Peoria	6,025	-4
	Rockford	6,830	-4
	Springfield	5,429	2
IN	Evansville	4,435	9
	Fort Wayne	6,205	1
	Indianapolis	5,699	2
	South Bend	6,439	1
IA	Burlington	6,114	-3
	Des Moines	6,588	-5
	Dubuque	7,376	-7

	City		
	Sioux City	6,951	-7
	Waterloo	7,320	-10
KS	Dodge City	4,986	5
	Goodland	6,141	0
	Topeka	5,182	4
	Wichita	4,620	7
KY	Covington	5,265	6
	Lexington	4,683	8
	Louisville	4,660	10
LA	Alexandria	1,921	27
	Baton Rouge	1,560	29
	Lake Charles	1,459	31
	New Orleans	1,385	33
	Shreveport	2,184	25
ME	Caribou	9,767	-13
	Portland	7,511	-1
MD	Baltimore	4,654	13
	Baltimore	4,111	17
	Frederick	5,087	12
MA	Boston	5,634	9
	Pittsfield	7,578	-3
	Worcester	6,969	4
MI	Alpena	8,506	-6
	Detroit (city)	6,232	6
	Escanaba	8,481	-7
	Flint	7,377	1
	Grand Rapids	6,894	5
	Lansing	6,909	1
	Marquette	8,393	-8
	Muskegon	6,696	6
	Sault Ste. Marie	9,048	-8
MN	Duluth	10,000	-16
	Minneapolis	8,382	-12
	Rochester	8,295	-12
MS	Jackson	2,239	25
	Meridian	2,289	23
	Vicksburg	2,041	26
MO	Columbia	5,046	4
	Kansas City	4,711	6
	St. Joseph	5,484	2
	St. Louis	4,900	6
	St. Louis	4,484	8
	Springfield	4,900	9
MT	Billings	7,049	-10
	Great Falls	7,750	-15
	Helena	8,129	-16
	Missoula	8,125	-6
NE	Grand Island	6,530	-3
	Lincoln	5,864	-2
	Norfolk	6,979	-4

	City		
	North Platte	6,684	-4
	Omaha	6,612	-3
	Scottsbluff	6,673	-3
NV	Elko	7,433	-2
	Ely	7,733	-4
	Las Vegas	2,709	28
	Reno	6,332	10
	Winnemucca	6,761	3
NH	Concord	7,383	-3
NJ	Atlantic City	4,812	13
	Newark	4,589	14
	Trenton	4,980	14
NM	Albuquerque	4,348	16
	Raton	6,228	1
	Roswell	3,793	18
	Silver City	3,705	10
NY	Albany	6,875	-1
	Albany	6,201	1
	Binghamton	7,286	1
	Buffalo	7,062	6
	NY(central park)	4,871	15
	NY(Kennedy	5,219	15
	NY (LaGuardia)	4,811	15
	Rochester	6,748	5
	Schenectady	6,650	1
	Syracuse	6,756	2
NC	Charlotte	3,181	22
	Greensboro	3,805	18
	Raleigh	3,393	20
	Winston-Salem	3,595	20
ND	Bismarck	8,851	-19
	Devils Lake	9,901	-21
	Fargo	9,226	18
	Williston	9,243	-21
OH	Akron-Canton	6,037	6
	Cincinnati	4,410	6
	Cleveland	6,351	5
	Columbus	5,660	5
	Dayton	5,622	4
	Mansfield	6,403	5
	Sandusky	5,796	6
	Toledo	6,494	1
	Youngstown	6,417	4
OK	Oklahoma City	3,725	13
	Tulsa	3,860	13
OR	Eugene	4,726	22
	Medford	5,008	23

	Portland	4,635	23
	Portland	4,109	24
	Salem	4,754	23
PA	Allentown	5,810	9
	Erie	6,451	9
	Harrisburg	5,251	11
	Philadelphia	5,144	14
	Pittsburgh	5,987	5
	Pittsburgh	5,053	7
	Reading	4,945	13
	Scranton	6,254	5
	Williamsport	5,934	7
RI	Providence	5,954	9
SC	Charleston	2,033	27
	Charleston	1,794	28
	Columbia	2,484	24
SD	Huron	8,223	-14
	Rapid City	7,345	7
	Sioux Falls	7,839	-11
TN	Bristol	4,143	14
	Chattanooga	3,254	18
	Knoxville	3,494	19
	Memphis	3,232	18
	Nashville	3,578	14
TX	Abilene	2,624	20
	Austin	1,711	28
	Dallas	2,363	22
	El Paso	2,700	24
	Houston	1,396	32
	Midland	2,591	21
	San Angelo	2,255	22
	San Antonio	1,546	30
	Waco	2,030	26
	Wichita Falls	2,832	18
UT	Salt Lake City	6,052	8
VT	Burlington	8,269	-7
VA	Lynchburg	4,166	16
	Norfolk	3,421	22
	Richmond	3,865	17
	Roanoke	4,150	16
WA	Olympia	5,236	22
	Seattle-Tacoma	5,145	26
	Seattle	4,424	27
	Spokane	6,655	2
WV	Charleston	4,476	11
	Elkins	5,675	6
	Huntington	4,446	10
	Parkersburg	4,754	11
WI	Green Bay	8,029	-9
	La Crosse	7,589	-9

	Madison	7,863	-7
	Milwaukee	7,635	-4
WY	Casper	7,410	-5
	Cheyenne	7,381	-1
	Lander	7,870	-11
	Sheridan	7,680	8

Definitions

Air Change: the amount of air that is required to completely replace the air in the boiler and associated flue passages.

Air, Primary: Air that mixes with the fuel to provide combustion.

Air, Secondary: Air that mixes with the flue gases to dilute the air going outside or to the chimney.

Air Separator: A device located in the supply pipe for a hydronic boiler that removes the entrained air from the water.

Air shutter: A device that controls the airflow to the burner

AWG: American Wire Gauge

Backflow Preventer: A device that will limit the backflow of boiler water into the potable water in a building or system.

Back Pressure: The pressure that is on the outlet of a steam trap.

Barometric Damper: A damper that is installed in the flue piping that will control the excessive draft in a category1 type boiler by introducing boiler room air.

Boiler: A closed vessel that heats water or creates steam

Boiler Design Temperature: It is the outside temperature at which the heating system can still provide heat to the building. It will be one of the coldest temperatures during an average winter.

Boiler, High Pressure: A boiler, which generates steam to pressures above 15 Psig

Boiler, Low Pressure: A boiler, which generates steam to pressures below 15 Psig

Boiler, Hydronic: A boiler, which heats water below the flash point.

Boiler, Cast Iron: A boiler, which uses cast iron as its heat exchanger. The cast iron boiler is sometimes available with the boiler sections disassembled and would require field erection.

Boiler, Copper: A boiler, which uses copper as its heat exchanger.

Boiler Feed Unit: A tank that gathers condensate from a steam system. All makeup water required for the steam system is introduced into the boiler feed tank.

Boiler, Steel: A boiler, which uses steel as its heat exchanger.

Boiler, Fire Tube: A steel boiler where the flue gases travel through the tubes inside the boiler. The heating medium, water or steam, surround the tubes.

Boiler, Water Tube: A steel boiler where the flue gases travel around the tubes inside the boiler. The heating medium, water or steam, travel through the tubes.

Boiler, Modular: A heating system consisting of several smaller boilers.

Breeching: A conduit that transports the combustion by products from the boiler to the outside or to the chimney. It is also called a flue.

Btu (British Thermal Unit): The amount of heat required to raise one pound of water, one degree F

Btuh: Btu's in one hour

Burner: A mechanical device that mixes air and fuel to provide ignition and combustion of the fuel.

160

Burner, Atmospheric: A burner that uses natural draft and gas pressure to provide combustion.

Burner, Power: A burner that uses an internal blower to mix the fuel and the air for combustion.

Carbon Dioxide: This is a gas that is produced as a by-product of combustion. It is also referred to as CO_2.

Carbon Monoxide: This deadly gas is odorless and tasteless. It is produced when there the combustion is out of adjustment. It is often referred to as CO.

Carboxyhemoglobin: This is formed in the blood when carbon monoxide and hemoglobin are combined due to the inhalation of carbon monoxide.

Compression Tank: A tank that is used in a hydronic system that will absorb the expansion of the water once it is heated. It is sometimes called an Expansion Tank.

Combination Low Water Cutoff/Pump Control: This control is mounted on a steam boiler to maintain the correct water level in the steam boiler. It has two sets of contacts inside. When the water levels drops to the level of the first setting, the control starts the boiler feed pump to replenish the water inside the boiler. If the water continues to drop, the second set of contacts set a lower elevation, cut the power to the burner.

Combustion Air: The air that is introduced from the outside that is required for the proper combustion of the fuel.

Combustion Analyzer: A device that measures the flue gas from a boiler and displays the different components. It will also display the efficiency of the boiler.

Condensate: Condensed water because of the removal of latent heat from steam.

Condensate Tank: A tank that collects condensate from the steam system. The condensate tank has an internal float that will energize a pump once the level rises to the set point of the float.

Condensing Boiler: A boiler that is designed to allow the flue temperatures to drop below the dew point temperature.

Control, Operating: A device that starts or stops the burner. This is usually set for a lower temperature or pressure than the Limit Control.

Control, Limit: A device that starts or stops the burner. This is usually set for a higher pressure or temperature than the operating control. In most applications, it has a manual reset feature.

Dew Point Temperature: The temperature at which warm humid air is cooled enough to allow the water vapor to condense into water.

Dirt Leg: A series of nipples and a pipe cap that are installed just before the train to capture any dirt that is in the gas line before it enters the gas train.

Draft: The pressure differential between atmospheric pressure and the pressure in the flue and boiler.

Draft Diverter: An air opening that introduces secondary air to the flue after the main combustion.

Draft, Mechanical: The pressure differential between atmospheric pressure and the pressure in the flue and boiler that is induced because of a fan or blower.

Draft, Natural: The pressure differential between atmospheric pressure and the pressure in the flue and boiler without a fan or blower.

Dual Fuel Burner: A burner, which has two fuel sources that it can use. It is usually natural gas and #2 fuel oil.

Emergency Door Switch: This manual switch is located at all exits from a boiler room that will shut off the boiler in the event that it is engaged.

Equalizing Line: The pipe that goes from the steam piping on a boiler to the bottom of the boiler. It will equalize the pressure in the boiler.

Expansion Tank: A tank that is used in a hydronic system that will absorb the expansion of the water once it is heated. It is sometimes called a Compression Tank.

Firing Rate: The burning rate of fuel and air in the burner.

Firing Rate Control: A control that senses the temperature or pressure of the heating system. It will regulate the burner between low and high fire to meet the desired set point. It is sometimes called the modulating control.

Flue: A conduit that transports the combustion by products from the boiler to the outside or to the chimney. It is also called a breeching.

Flue Gases: These byproducts of combustion are produced by the burner. They will be vented from the boiler with a flue.

Fuel Train: A series of components, including gas pressure regulator and gas valves, that are located in the gas piping directly attached to the burner. This is also called a gas train.

Gas Pressure Regulator: A device that controls the gas pressure supplied to the burner.

Gas Pressure Switch: A safety device that senses the available gas pressure and will shut the boiler off in the event that the pressure is outside of the setting. There are usually two types of gas pressure switches on a boiler. The High Gas Pressure Switch that is located in the gas train downstream of the gas pressure regulator and the electric gas valves. It will shut the boiler off if the gas pressure is higher than the setting. The Low Gas Pressure Switch is located downstream of the main gas pressure regulator. It will shut the boiler off if the gas pressure is below the setting.

GPM: Gallons per Minute

Heat, Latent: The amount of heat required to cause a change of state.

Heat, Sensible: The amount of heat required to cause a change in temperature

Heating Medium: The material that the boiler heats. It could be steam, water or some other type of fluid.

High Fire: This is the highest design firing rate of the burner. It is the 100% firing rate.

Hydronic System: A heating system that uses water as the heating medium instead of steam.

Lag Boiler: The boiler that is not the first boiler to start when there is a call for heat.

Lead Boiler: The boiler that is the first boiler to start on a call for heat.

Life Cycle Cost: This is the amount of money that the system costs the owner over the estimated life of the unit. It will include fuel and repair costs as well estimated repair parts.

Lockout: A safety shutdown that requires a manual reset of the control or safety device.

Low Fire: This is the lowest design firing rate of the burner.

Low Fire Start: This switch verifies that the burner is in the "Low Fire" position before opening fuel valves.

Low High Low Fire: A burner that starts at low fire and then goes to high fire if there is still a call for heat. As the temperature or pressure gets close to the set point on the firing rate control, the burner will drop to low fire.

Low High Off Fire: A burner that starts at low fire and then goes to high fire if there is still a call for heat. The burner will stay at high fire until the call for heat has ceased.

Low Water Cutoff: A device that senses the water level inside the boiler and will shut down the burner if the water level drops to an unsafe level.

Modulating Burner: A burner that will operate at any position from low to high fire to meet the demands of the modulating control.

Modulating Control: A control that senses the heating medium and will send a signal to the burner that will set the burner at any position from low to high fire.

Non-Condensing Boiler: A boiler that is designed to keep the flue temperatures above the dew point temperature.

Pilot, Continuous: It is a pilot flame that burns all the time, regardless of whether the burner is firing.

Pilot, Intermittent. It is a pilot that lights when there is a call for heat. The pilot will stay light during the entire time that the main burner is firing.

Pilot, Interrupted: It is a pilot that lights when there is a call for heat. The pilot will shut off once the main flame is established.

Pot Feeder: A device that is used to introduce water treatment into a heating system.
PPM: Part per Million.

Prepurge: On a call for heat, the burner blower starts to purge the boiler combustion chamber and flue passages of any unburnt fuels. It will operate for a duration long enough to provide several air changes inside the boiler.

Proprietary Parts: parts that are only available from the manufacturer or have limited distribution.

Post purge: The burner blower will operate for a time after the call for heat has been satisfied to purge any unburnt fuel.

Relief Valve: A valve located on a boiler that will relieve the internal boiler pressure if the pressure rises to the rating of the relief valve.

Reset Control: A control that will lower the supply temperature in a hydronic system as the outside temperature increases.

Sidewall Venting: Boiler flue that is piped to the sidewall of the building rather than a chimney or stack.

Spill Switch: A device that is located by a draft diverter or a barometric damper that senses rollout of the flue gases and shuts off the burner.

Swing Joint: A series of piping that helps control the piping expansion on a cast iron boiler in a steam system.

Siphon: A piece of pipe that traps water inside to form a water seal to protect the control. It is also called a pigtail.

Velocity, Steam: The speed at which steam travels in a system.

Index

22071492R00090

Made in the USA
Charleston, SC
09 September 2013